how
to be
sad

how
to be
sad

*Everything I've Learned About
Getting Happier by Being Sad*

HELEN RUSSELL

HarperOne
An Imprint of HarperCollinsPublishers

HarperCollins books may be purchased for educational,
business, or sales promotional use. For information, please email
the Special Markets Department at SPsales@harpercollins.com.

Originally published as *How to Be Sad* in Great Britain in 2021
by 4th Estate, an imprint of HarperCollins Publishers UK.

FIRST HARPERCOLLINS PAPERBACK PUBLISHED IN 2022

Library of Congress Cataloging-in-Publication Data
is available upon request.

ISBN 978-0-06-311536-1

22 23 24 25 26   LSC   10 9 8 7 6 5 4 3 2 1

If you picked this up: it's for you.
I wrote it for you (and also my mom).

Contents

CONTENTS

Introduction

We're eating oranges in the sunshine. Sitting cross-legged on freshly cut grass, in a cemetery. The woman next to me is wearing a red beret and weeping. This isn't our usual meeting place—normally there'd be baked goods and a coffee frother involved. But today, my mother and I have made a pilgrimage, to be dwarfed by giant fir trees, feel the sun on our skin and a sadness deep within. It's not most people's idea of a fun day out, but it is important—I know. Because having spent the past eight years researching into happiness worldwide, I've inadvertently become something of a specialist in sadness.

I began to notice that many of the people I met were so obsessed with the pursuit of happiness that they were phobic about feeling sad. I'd speak to people who had just lost loved ones who would ask how they could be happy. I'd meet people who'd recently been made redundant. Or had become homeless. Or had a bad breakup. Or those with caring responsibilities who weren't being properly looked after themselves. Or people experiencing truly terrible things who still asked: "So why aren't

ix

I happy?" I would try to explain that, sometimes, we need to be sad. How sadness is what we're supposed to feel after a loss and how sorrow is the sane response when sad things happen. But a lot of us are conditioned to be so averse to "negative emotions" that we don't even recognize them, much less acknowledge them or give ourselves permission to feel and process them. I've lost count of the number of people who've said, "I just want to be happy," at times when this is almost impossible. When we lose a job, or a home, or a relationship, or a family member or *anything*—it's normal to be sad.

Sadness is defined as the natural response to emotional pain, feelings of loss, helplessness, hopelessness or disappointment. Sadness is normal. It's also inevitable. As Desmond Tutu said: "I am sorry to say that suffering is not optional." Or, in the words of Westley from *The Princess Bride*: "Life is pain . . . Anyone who says differently is selling something."

Sadness happens, to all of us—sometimes in heartbreakingly awful ways. Only, in much of the world, we're pretty poor at handling it. This can be isolating for those experiencing it and baffling for those trying to help loved ones through tough times.

There will always be tragedy. There will always be loss. And there'll be day-to-day troubles, too—from the mundane, to the dispiriting and demotivating. Sadness is far more layered and complex than happiness and it's everywhere. We can't avoid sadness. But we can learn to handle it better. And it's high time we started talking about it. Because the current approach to so-called negative emotions isn't working—and sadness can actually be helpful. As the Danish philosopher Kierkegaard writes: there is "bliss in melancholy and sadness" and researchers from the University of New South Wales have found that accepting and allowing for temporary sadness helps to improve our attention to detail, increases perseverance, promotes generosity and even makes us more grateful for what we've got.[1] Sadness has a point. It can tell us when something is wrong, if we let it. Sadness is the temporary emotion that we all feel on occasions when we've been hurt or something is wrong in our lives. It is a *message*.

We depend on each other to survive as a species and sadness is the emotion that makes us remember this[2]—because the most common ways to avoid sadness are really ways to avoid *feeling*. Like trying not to get too close to someone for fear of getting hurt [*I've been there . . .*]. Or avoiding the pursuit of meaningful goals in case we "fail" at them [*check . . .*]. Or forming addictions to blot out the pain or numb our senses and so "protect" ourselves [*waves*]. Or working all the time, busying ourselves with the hamster wheel of life and distracting ourselves from uncomfortable feelings by scrolling through social media, just for example . . . [*is it hot in here?*]. If we aim to avoid sadness, even a little, we limit our existence and put ourselves at greater risk of normal sadness tipping over into something more serious.

Suppressing negative or depressive thoughts, to the extent that many of us probably do on a daily basis, has been proven to backfire spectacularly, resulting in depressive symptoms, according to studies. The Harvard University psychologist Daniel Wegner famously led a thought experiment in 1987 where subjects were told *not* to think about white bears,[3] inspired by the Russian writer Dostoevsky, who once wrote: *"Try to pose for yourself this task: not to think of a polar bear, and you will see that the cursed thing will come to mind every minute."*[4]

So Wegner decided to put this idea to the test.

For five minutes, participants were asked *not* to think about a white bear, but to ring a bell each time a white bear crossed their minds. Participants in a second group were allowed to think of anything they wanted, but continued to ring a bell each time the thought of a white bear surfaced. The second "expression" group rang the bell far less frequently than the first group, who'd been suppressing their thoughts. A second experiment replicated these findings and Wegner later collaborated with the psychologist Richard Wenzlaff to test the theory further, confirming that trying not to think or feel something sad makes us more prone to anxiety, depressive thoughts and symptoms.[5] It may sound counterintuitive, but Wegner and Wenzlaff concluded that fighting "sad" actually makes the "sad" worse.

This was certainly my experience.

I know "sad" just as intimately as I know "happiness" and this book isn't purely professional for me. My first ever memory is the day that my sister died, of sudden infant death syndrome (SIDS). My parents split up soon after. I've had a sketchy relationship with my body and with food. I've had career setbacks as well as relationships that have imploded in agonizing ways. Infertility, IVF and bed rest weren't exactly a hoot either. Even the things that have brought the greatest joy have also brought challenges. Challenges that have been harder to process than they might have been—because we don't tend to talk about sadness enough in our culture. Many of us will have been raised with the assumption that "what we don't talk about can't hurt us," and, for a long time, "not talking" about being sad was seen as a sign of strength. But really, the opposite is true. And learning how to be sad, better, is something we need right now, more than ever.

At the time of writing, we are still in the grip of a global pandemic, with hope that things will have improved by the time this book publishes. In the time of the coronavirus, much of what we relied on and took for granted was suddenly gone. The enforced slowness of lockdown stripped everything back, so that our internal dialogue could be heard more loudly—with no respite or escape to the busyness of "normal" life. Some have been separated from loved ones. Some are alone. Some are afraid. Some have been trapped at home in a relationship with someone with whom, it is now apparent, a continued relationship depended on *not* being trapped at home together. No one really knows what the world will look like moving forward or how we'll recover—

economically or emotionally, with high unemployment and a recession looming. Many of us will have experienced losses and all of us will have felt *something* change. And because we're more globally connected—at least, digitally, today—we're more aware of what's going on around us.

The attacks on trans rights and the Black Lives Matter movement further remind us that there is plenty to be sad about. COVID-19 has disproportionately impacted Black communities, both in terms of infection rates and higher mortality, as well as legal, social and economic inequalities.[6] This has led to a spike in anxiety and depression among African Americans.[7] We're experiencing record levels of sadness, worry and anger, globally, according to a recent annual Gallup survey.[8]

It's now estimated by the World Health Organization (WHO) that 264 million people globally are affected by depression.[9] Of course, sadness isn't the same as depression (I should know: spoiler alert, I've done both). WHO defines depression as *persistent* sadness and a lack of interest or pleasure in previously rewarding or enjoyable activities. Depression often disturbs sleep and appetite, resulting in poor concentration.

There are six common types of depression: the first is major depression, what many of us probably think of when we hear the word "depression"—a clinical condition with symptoms along the lines of the WHO's definition. Then there's persistent depressive disorder, referring to low mood lasting for at least two years but that may not reach the intensity of major depression. There's also bipolar disorder; seasonal affective disorder; a severe form of premenstrual syndrome called PMDD; and perinatal depression that can occur during pregnancy or in the first year after a baby is born (also known as postpartum depression).[10]

Clinical depression is serious and usually requires professional help.[11] But a refusal to sit with standard, unavoidable sadness—and a lack of knowledge as to how best to handle it—can lead to depressive symptoms (see Wenzlaff and Wegner).

Because sadness is "normal."

"Many people nowadays assume that if they're not happy, they must be depressed," says the philosopher Peg O'Connor, chair of the Department of Philosophy at Gustavus Adolphus College in Minnesota, who I will speak to for Chapter 5 of this book. "But life isn't like that—there's a whole swatch of emotions and ways of being that are viable. As Aristotle says, happiness is an ongoing activity; it doesn't mean that you're never unhappy or that hard things haven't happened. Life *is* hard and there *are* challenges—but that doesn't mean that you can't have a good life." I talk to Meik Wiking, CEO of the Happiness Research Institute in Denmark, about this and he tells me: "It's important that us happiness researchers underline that no one is happy all the time. That sadness is part of the human experience called 'life.'"

Much of what makes us sad may be unforeseen; few of us predicted the events of 2020 when COVID-19 knocked us sideways. But other episodes of sadness can be relied on—scheduled, even. Researchers have found that our lives typically follow a u-shaped curve,[12] whereby we're happier at either end with a distinct dip in middle age. Economists David Blanchflower and Andrew Oswald started noticing recurring patterns in life satisfaction studies back in the 1990s. By 2017, they published meta-analysis showing conclusively that contentment declines for the first two decades of adulthood, hitting rock bottom in our forties, then creeping up again and sending us near giddy into our senior years. As unlikely as it sounds for those not at all happy to be nearing death, the trend plays out worldwide. The loss of happiness between the ages of twenty-five and forty has been found to be equivalent to one-third of the effect of involuntary unemployment.[13]

It was originally assumed that this dip came down to the burdens of middle age—like job stress, money worries and caring responsibilities in both directions. But then scientists realized that the same trend plays out in chimpanzees.[14] This suggests that the pattern is rooted in biological or even evolutionary factors, rather than the pressure of a large mortgage. One theory is that we and our chimp cousins need higher levels of well-being during life stages where we have fewer resources, like youth or old age. Another thesis is that as we age and time horizons grow shorter, we invest in what's most important in life—like relationships—and so derive increasingly greater pleasure from them.[15] In other words, we stop chasing fame/Ferraris/big bananas and learn to appreciate hanging out with our family instead. This is a nice idea, but basically no one's sure. The exact scientific explanation for the u-shaped curve currently remains unknown[16] (get used to this: brain science is very much a work in progress). What's certain is that we all experience painful life events and we all endure periods of sadness.

These should be times when we feel the most connection with our fellow human beings, yet they're often the times when we feel most alone—retreating rather than reaching out. We may be embarrassed to admit that we're sad. Other people may be embarrassed that we're sad. We may be embarrassed that *they're* embarrassed. Either way, we feel shame (more in Chapter 7).

Many of us instinctively tell ourselves that we "shouldn't" feel sad, that others have it worse. We worry that our sadness—our own brand of hurt—is somehow less "legitimate" than others'. "Unworthy" even. But pain is pain, however privileged. This is not to minimize or make light of others' suffering; it's about being aware and attentive to our own, too. We have to care about the world around us and help others. But we still get to hurt. If we're feeling sad, we should let ourselves feel sad and allow the emotion to move through us, since sadness can be useful and we all get sad to varying degrees.

No one can be happy all the time. The lows help us appreciate the highs and to be truly content, we have to make friends with "sad," too.

I've spent the past four decades learning how: in loss, heartbreak, friendship and family life; through addiction, adverse circumstances and in depression. The examples of "sad" in this book aren't comprehensive. We may not have experienced exactly the same "sad," but the steps are the same and by sharing my story, I hope I can encourage others to do the same. We are all different, but there is universality in the specifics.

For this book, I have consulted experts in their fields—from psychology to neuropharmacology, grief counseling to genetics, psychotherapy to neuroscience and doctors to dietitians. All are currently tooling up to help us in the fallout from coronavirus under the agreed assumption that we're all going to need to get better at being sad as a society. Starting now.

I've also been inspired and enlightened by fellow sad aficionados: comics, writers, explorers, idols, friends, family, the honest and the brave. By various characters who have intersected my life: from the people I have been watching on TV to those I've been reading or listening to who have shared their stories and advice for how to be sad, well. Loss comes up a lot, so I'll talk about "grief" as a universal term to express loss. Bereavement and mourning are more specific to death, but grief can refer to living losses, too—and we've all experienced those.

This is a book to show you how to be sad, well. A book of shared experiences and inspiration to help all of us realize that we are not alone. Because sadness is going to happen, so we might as well know how to "do it" right. And we can all get happier—by learning how to be sad, better.

Part I

How to Look After Ourselves When We're Sad

On loss and learning to accept it; the physical mani-festations of sadness; the perils of perfectionism; and why getting mad can help sometimes.

How boys *do* cry (everyone should); why abandonment and adolescence make for a heady combination; the truth about mangled hearts and the myth of the "just world."

What we can learn from anxiety and addiction; how grief is seldom neat; unemployment; injustice; and why we all need to practice more humanity.

Introducing my mom and dad, Orange Backpack Man, Kevin the driving instructor, the tall guy and T.

Featuring Professor Peg O'Connor, Professor Nathaniel Herr, psychotherapist Julia Samuel, the "Tear Professor" Ad Vingerhoets, Harvard University lecturer Dr. Tal Ben-Shahar, the Danish philosopher Søren Kierkegaard and Phil Collins.

1

Don't Fight It

I^T's 1983, IT's RAINING and Phil Collins's "You Can't Hurry Love" is playing on the radio. I won't know what irony is for at least fifteen years but this already seems cruel. Because it turns out you can't slow love down, either. I'm on the couch playing with my blue-haired doll when I hear the familiar squeak of hand on banister. It is my dad and he's carrying a suitcase. He's wearing bell-bottoms and a shirt with sleeves rolled up to his elbows, despite the fact that it's January. His hair is long, spilling over the edge of his collar. And it's brown. It's the early eighties, so most things are brown—the clothes, the decor, my hair and that of my parents. I am three years old and it's only been three months since a Very Sad Thing happened to our family, on October 31, 1982. Halloween, in fact. A day that changed all of us but that will remain unspoken of for years to come.

My dad's eyes used to crinkle when he smiled and my mom used to be "chatty," but now my dad doesn't smile and my mother doesn't chat. Nothing has been right since the Very Sad Thing happened, and now my dad is leaving.

He's back after a few days the following weekend but he doesn't stay the night. I know it's the weekend as I'm being allowed to watch TV in my pajamas after breakfast rather than having my hair and teeth brushed immediately afterward. This is strange. What's stranger still is that when my paternal grandparents visit, no one mentions the lack of sleepovers.

"You haven't told them?" I hear my mother whisper to my father in the kitchen.

Told them what?

My dad starts picking me up every Saturday and driving me to a popular chain of family-friendly eateries. I eat a lot of corn from the salad buffet then stuff my face with ice cream for dessert, so I don't have to talk too much but then feel sick after. My dad has started wearing a leather jacket and now drives a convertible Golf GTI. This would probably be termed a midlife crisis for most men, but my dad is only twenty-seven. So perhaps it's just a "crisis." I don't much like the convertible Golf GTI because having the roof down makes my hair blow about until I can't see and then get carsick and vomit. This makes the car smell so bad that having the roof down becomes a necessity; otherwise my dad will vomit, too. Nausea is fast becoming a constant.

These outings are . . . fine. But soon our weekly lunchtime jaunts morph into monthly, overnight expeditions. My dad is staying in a small apartment in London with his "new girlfriend," her sister and her teenage son. There isn't really room for all of us so I share a bunk bed with the fourteen-year-old boy. Sunday mornings now start with a teenage boy swinging his legs down from the top bunk and scratching his ass through boxer shorts. It's confusing and it doesn't smell great. But then, nothing seems to smell great anymore.

My mom and I move to be closer to my grandmother, a formidable woman who looks like a cross between the Queen and Margaret

Thatcher. I start preschool in September while my mother goes back to work. No one tells the school what has been happening in our family until my mother is confronted with a picture I have drawn that my teacher is particularly pleased with—a drawing of my mom, my dad, my baby sister and me. My mom turns white at this and has to explain to the teacher that my baby sister is "no longer with us" and that my dad isn't coming back, either. I am baffled by this.

Dad gone too?

To cheer us up after this upsetting incident, we decide that it is my blue-haired doll's birthday and my mom bakes her a cake. I have little appetite but stuff it down regardless. I turn out to be pretty good at this. Food is a way of showing love—and who can be sad eating cake? Sadness can also, I learn, be resisted or at least *rescheduled* by eating cookies, white bread and cereal straight from the package. All hail carbohydrates.

My dad and his new girlfriend want a place of their own to live, but they haven't got enough money to do this "what with me to pay for as well" and so my dad becomes stressed. He also becomes forgetful.

Pigtailed. Wide-eyed. Aged five. I remember waiting. Sitting on the beige-carpeted bottom step of my mother's new house with a packed suitcase by my side. Toothbrush, pajamas, two changes of underwear (just in case), my favorite purple sweater and brown corduroy pants (the 1980s . . .) have been stowed away with care. But the blue-haired doll has been left out "for air" and is held tight in my arms. The clock shows both hands pointing upward, right at the top. This is the time my mom said that my dad would come. I have been "good," so he will come. He has to come. So I wait. And then I wait some more. Minutes tick by, audibly, until the big hand on the clock points toward the floor. The clock is now making an altogether different shape to the one my mother and I have drawn together on pieces of paper "to practice." My mother's voice becomes slightly higher as she assures me: "Everything's fine!" over and over. She alternates between checking the street for signs of life, trying the telephone, and even, unusually, offering the option of cartoons. But I will not be moved. I sit, eyes trained on the front door, for three hours.

He does not come.

When my baby sister was here, my dad did not forget things, and life was okay. Now it's just me, my dad is increasingly forgetful and life is very much *not* okay. This confirms a new, niggling fear that has begun to develop: that it would probably have been better if I'd gone instead and it's all my fault that Dad left.

I'm not special: preschoolers typically believe that they're responsible for their parents' separation. "What you have there is a case of childhood omnipotence," the psychologist Aphrodite Matsakis tells me three decades later. This is a well-documented tendency of some children (and some adults) to think that the world revolves around them and that they control everything that happens in it. "Some young children have trouble seeing things from others' point of view and tend to think that they're the center of everything—as well as the *cause* of everything. They often think that if they wish something it might come true. It's an exaggerated sense of responsibility based on the belief that 'I, personally, have the duty and power to save loved ones in trouble.'"

No one tells me otherwise. No one tells me much at all. So I make it up. "If we don't tell them the truth, children do make it up," says Jane Elfer, a child and adolescent psychotherapist at a large London hospital. "They invent their own version of what's happened—their own reality or faulty ideas. Often, what children imagine is even worse than what's actually occurred," she says, "so from a really young age, we need clear, concrete and specific communication to avoid misunderstandings. We have to get better at unhappiness as a society—if something sad has happened you need to allow and accept this."

We do not accept it: we fight it. Ignore it, even.

The paperwork comes through and my parents are officially divorced. Despite the commonly held myth that most couples split up after the death of a child, around 72 percent of parents who are married at the time of their child's death remain married to the same person.[1] It is doubtless insanely painful, and cracks in a relationship turn to chasms under pressure. But this doesn't necessarily mean that we or our relationships are broken (although it may feel that way). According to the American Psychological Association, 40 to 50 percent of married couples in the United States end up divorced.[2] So bereaved couples may actually be *more* likely to stay together—and one loss doesn't necessarily have to lead to another. Grief is the price we pay for love, but if we're not prepared for this and we've been raised to demand happiness, or at least a numbing out of pain at every turn, we're less able to ride out the storm. If we expect too much of ourselves and our relationships after a loss, we will be disappointed. I fully understand the impulse to run for the hills in an attempt to "escape" sadness and pain: most of us have been raised to run for the hills. No judgment toward anyone whose longest relationship to date is with "the hills." Really, I get it (*I heart hills . . .*). People do crazy things. Neither of my parents were saints. And divorce is often the best course of action for both parties. But it's worthwhile to remember that there is another way. When we're experiencing loss, from low-level sadness to the catastrophic, life-changing kind, we *will* feel bad—that's normal. If we learned to accept that things were going to be hard, we might be better equipped to endure periods of extreme sadness. Something I wish someone had told my family in the 1980s. But they don't. Because no one tells each other anything.

Instead, I join the esteemed legions of men and women worldwide with "Daddy Issues." I grow up with a single parent who does the work of two—a woman who, fortunately for me, is extraordinary in her strength and resilience. There are some pros to being the child of a single mother: I will grow up blissfully ignorant of the gendered nature of many domestic tasks, since in my house what needs to be done gets done, by *her*. I will become, like my mother, excellent in a crisis. I will value independence, although unfortunately, to the extent that I will become hooked on the stuff, wary of commitment or going all-in with anyone (I've seen where *that* can get you). I will insist on "room to breathe" in every relationship I'll ever have. I will struggle to negotiate—there was no need in our house, since one person made all the decisions. And I will see how keeping busy is a way to keep going. To fight the pain. Ish. The world already makes no sense to me so I make sense of it myself. I'm told, regularly, not to be sad and not to cry. So I don't. No one does. Until the urge to cry or "feel sad" becomes strangely unfamiliar. *Alien*, even.

The late psychologist Haim Ginott wrote in *Between Parent and Child*[3] that: "Many people have been educated out of knowing what their feelings are. When they hated, they were told it was only dislike. When they were afraid, they were told there was nothing to be afraid of. When they felt pain, they were advised to be brave and smile." Children look to parents for how to regulate their own emotions because they don't yet know how to do it themselves. But if caregivers don't know either, or were never taught because the "bad feelings" were anesthetized away, then we're really in trouble. And trying to fight "sad" is something many of us are taught from birth. We administer sweet-tasting painkillers to babies after their first vaccinations, or anytime they have to travel, or feel discomfort from teething. The message is clear: being a good caregiver means never allowing your child to suffer, no matter what the reason for this might be. We live in a culture where distress demands to be alleviated and sadness is supposed to be "solved" rather than experienced.

"With most things in our life these days if we have a problem, we expect technology or medicine to 'fix us,'" says the psychotherapist and grief expert Julia Samuel, MBE, "but sadness doesn't work like this. We are not taught to experience a little bit of pain so that we can learn to deal with big pain."

We try to fight it: to lessen discomfort as a society, almost on auto-pilot. Only by doing so, we're all worse off.[4] Professor Nathaniel Herr from American University in Washington, DC, an expert in emotional regulation, says, "'Sadness' is *really* important. People need to recognize it and what it provides. I have people saying to me: 'I just don't want to feel anxious anymore—I don't want to feel "sad,"' and I have to say to them: 'I can't help you with that!' Because you shouldn't want '*not* to feel sad.'" This is something that even Herr's psychology students have a tough time getting their heads around. "If I ask them, 'Why do we have sadness?' most will say something like, 'Well, we couldn't have happiness if we didn't have sadness! It's like light and shade.' But that's not it: they're ignoring the function that sadness has socially. It sends out a signal, like, 'Hey! Come help me!' to make other people rally around." Herr also takes the view that we're often sad when we're stuck somehow and don't know how to get ourselves out of a situation, "which makes sadness profoundly useful."

"Sadness is a problem-solving type of emotion," he says. "It produces rumination. I see rumination as the cognitive manifestation of the emotion 'sadness'—just like worry is the cognitive manifestation of the emotion 'anxiety.'" So sadness is an emotion that's important for making us stop and consider where we are, before we can move on to the next stage in life.

This is an idea originally flung around by the Danish philosopher Søren Kierkegaard. He argued that sadness and despair were not only "bliss"-inducing and inevitable: they were also necessary for change. I visit the Kierkegaard expert and psychology professor Henrik Høgh-Olesen at Aarhus University in Denmark, who tells me: "Kierkegaard is *all* about despair—and we need despair. When you feel that sadness or emptiness or anxiety—that existential feeling that makes you stop and wonder—that's an opportunity to make a change in life and row your boat against the stream."

Tide?

"*Stream*," he asserts. Psychologist + lecturer + expert on a notoriously tricky nineteenth-century philosopher proves to be a uniquely authoritarian combination so I don't quibble.

"You need these heavy feelings to help you navigate life." He thumps a tanned wrist on his desk. Sitting under ceiling tiles surrounded by potted plants, I'm transported back to my own student days and a late-essay-scolding in a professor's office circa 1998, experiencing a sense of déjà vu. "Sadness and despair give you purpose. We should be guided by these feelings: to think, What am I doing with my life?"

I feel small and wonder what, indeed, I am doing with my life.

And, er, what happens if we don't let ourselves be guided by despair?

"Then?" His voice rises. "Then? Then you're a robot! Just a robot who eats, sleeps and excretes!"

Right. Good. So "despair" is necessary for change and sadness is necessary for life otherwise we're all just shitting robots. *Got it*. So we need to stop fighting these "negative emotions" and start feeling them. Because if we don't, the consequences can be dire.

"If we don't accept and process sadness, it can manifest physically," warns the child and adolescent psychotherapist Jane Elfer. She explains that illness—actual, real illness—may be the only way a child can express their feelings. "They will have unexplained tummy aches or headaches, for example—and then of course there's the psychological impact," she says, "especially when it comes to loss."

My gut becomes a barometer of my mental state as a child but I have a lot of trouble identifying whether my tensed-up innards are because of hunger, tiredness, stress or sorrow. Since eating is a quicker fix than overcoming stress or sorrow—or napping—I usually feed the tightening cords in my stomach as a first response, just in case. This is fairly common. The late psychoanalyst Joyce McDougall wrote about how grief can manifest via loss of appetite, or an increase in appetite in an attempt to "fill the void." "Little children don't always know which bit of them is hurting, at least not before they're around ten years old," says Ross Cormack, psychotherapist and lead practitioner at Winston's Wish, the childhood bereavement charity, "and sadness is often felt in the stomach." Sad children also tend to be more panicky and live on cortisol—the stress hormone—and adrenaline, putting them in a constant state of fight or flight. Food can do a passable job of dampening this down and subduing us. But it's a shortsighted solution (and I say this as someone who once ate a whole loaf of white bread, just to check). It's only a "fix" until the next wave of "feelings" that need processing.

Other common physical symptoms of grief[5,6] include a tightness of the chest or throat, oversensitivity to noise, difficulty breathing, feeling very tired and weak, a dry mouth, an increase or decrease in appetite, difficulty or fear of sleeping, as well as aches and pains.[7] A 2014 study even found that older adults experiencing grief are less likely to produce some types of white blood cells, leaving them more prone to infections.[8] Unresolved grief is said to cause 15 percent of all psychological disorders, according to Julia Samuel in her book *Grief Works*.[9] Samuel also notes that bereaved children are more susceptible to addiction and mental health problems in later life if they haven't properly dealt with their emotions.

Those of us who struggle to regulate our emotions, Herr explains, usually experience this in three ways. The first is in terms of sensitivity to cues ("getting the emotions sooner than others might"). The second is intensity ("getting the emotions at a more intense level"). And the third is taking longer to return to a baseline or "normal." "It's up to adults," says Herr, "parents, usually, to give appropriate feedback to children so that they learn this in childhood. So if a kid says, 'I'm sad,' and a parent says, 'No, you're fine!' that wouldn't be helpful. Parents need to accept the emotion that a child is having and help the child to label it for the child to learn to identify that emotion, handle it, and not feel shame or confusion about it." Because, he says, every emotion is useful in its place: "If we just learned to accept and tolerate a full range of emotions to start with—especially the 'negative' ones— we'd all be better off." Especially, apparently, in the US. Researchers have found that, culturally, Americans are "outliers" in terms of their desire to minimize negative emotions.

Psychologist Jeanne Tsai from Stanford University's Culture and Emotion Lab has found that an obsession with the pursuit of happiness has led many Americans to view sadness as a "failure" and something that is the individual's responsibility.[10] As the daughter of Taiwanese immigrants raised in the US, Tsai became interested in how American attitudes differed from those typically found in East Asian cultures. "In the US I observed a real emphasis on wanting to feel happiness and avoid sadness, at all costs—far more so than in other cultures," she tells me when I get in touch. In East Asia, by contrast, the concept of negative feelings is rooted in Buddhist, Taoist and Confucian traditions and viewed as "situationally based" or circumstantial.[11] This means that individuals don't bear the weight of their negative experiences alone, "And negative feelings or experiences can even *foster* social ties in East Asian culture," says Tsai. In East Asia, negative emotions are seen as "inevitable and transient elements of a natural cycle"—or part of life rather than something to be feared as a risk to our mental or even physical health.

We've all seen studies saying "happy" people are healthier—and we certainly spend enough time and money trying to be happy in the

West. I used to believe in this myself. I spent years devouring and dutifully repeating research that "proved" happier people were healthier, ergo we should strive for happiness at all costs. But this isn't the whole story. Because in cultures where being sad is seen as "okay," sadness has been shown to have far less of a negative impact on health. "Researchers have analyzed the differences in approach to negative feelings and health in Japan and the US—a good comparison because both are modernized, democratized, industrialized societies with well-developed systems of healthcare," says Tsai. But these societies have very different ideas about negative emotions. As one Japanese psychiatrist told the Association for Psychological Science: "Melancholia, sensitivity, fragility—these are not negative things in a Japanese context. It never occurred to us that we should try to remove them, because it never occurred to us that they were bad."[12] Unlike in the US, where "sad" is empirically viewed as "bad." And it's this perception of sadness that can make us sick.

In the US, lower positive emotions are linked with higher BMI and less healthy blood lipid profiles (important indicators of health). But in Japan, studies show that people with lower positive emotions are pretty much . . . fine. So emotions have a different impact on our health depending on our culture; and being sad only makes us sick if we're terrified of being sad.

Another study from the University of California, Berkeley, found that people who accept rather than judge their mental experiences have better health outcomes.[13] Those who avoided their negative feelings or judged themselves harshly for feeling bad were more likely to report mood disorders and distress. Because if we view "sadness" as something "wrong" or even "abnormal," we're more prone to pathologize it.

In *The Loss of Sadness: How Psychiatry Transformed Normal Sorrow into Depressive Disorder*[14] (the title says it all), sociology professors Allan

V. Horwitz and Jerome C. Wakefield argue that the massive rise in depression in recent years has less to do with the pressures of modern life, and more to do with overdiagnosis. The medical historian Edward Shorter argues that psychiatry's "love affair" with the diagnosis of depression has become a death grip, asserting that most patients who get the diagnosis of depression are also anxious, fatigued, unable to sleep and have all kinds of physical symptoms.[15] Horwitz, Wakefield, Shorter and others suspect that many of us are being diagnosed as depressed when we are, in fact, "sad"—a direct result of a definition in a single, albeit significant, book.

The American Psychiatric Association's *Diagnostic and Statistical Manual of Mental Disorders* (the *DSM*) is a weighty tome used to diagnose all things mind-related in the US. The first edition was published in 1952 in an attempt to unify mental health approaches, but when it came to "major depressive disorders" the *DSM* focused on symptoms rather than context. This meant that the distinction between "actual medical issue" and "ordinary sorrow" was done away with. Anyone exhibiting five or more[16] "symptoms" for two weeks could be diagnosed with clinical depression—even if their low mood, decreased appetite or poor sleep and so on. had a thoroughly understandable explanation, such as heartbreak or financial worries. Earlier editions of the *DSM* included a "grief clause," stating that people couldn't be diagnosed with depression within two months of bereavement. But the latest version published in 2013 (*DSM-5*) scrapped this, doing away with the distinction between understandable sadness and medical condition. Supporters of the *DSM-5*'s decision argue that grief is a common precursor of depression and given the serious risks of unrecognized major depression, removing the bereavement exclusion was a reasonable decision. But it also means that responses to grief now can be labeled as pathological disorders, rather than being recognized as normal human experiences.

Psychologists in the rest of the world are supposed to use the World Health Organization's *International Classification of Diseases and Related Health Problems (ICD)* as a reference. But the *DSM* remains enormously influential.[17] So now the whole world is taking a leaf out of the US mental health playbook. A problem, since "Americans really don't like to be 'sad,'" as Tsai puts it—a proclivity she puts down to "frontier values."

"The first settlers from Europe were a self-selecting, intrepid group," says Tsai. "People who anticipated positive outcomes, were willing to take risks and who handled negative feelings or situations by *leaving them* in the hope of something better." For early pioneers, overcoming hardship was seen as a virtue, whereas wallowing in adverse circumstances was not. Consequently, the American approach to mental health today tends to be fervently forward-facing. One of the most popular modes of treatment is cognitive behavioral therapy (CBT), a forward-facing "back on your feet" intervention that seeks to change negative patterns of thinking. Many of CBT's pioneers came from the US[18] and whereas European psychologists are frequently influenced by Freud's "blame your father" backward-facing stance, Americans tend to prefer the promise of a sorrow-free tomorrow. Only seeing sadness as transgressive—a "problem" to be medicated away—leaves us poorly equipped to manage the next time it comes calling. Pathologizing sadness sends a message that "discomfort" cannot—and should not—be tolerated.

In these days of supposed enlightenment, we may be "officially" on board with public displays of emotion (more in Chapter 11) and we're certainly inundated with reality shows that end with a Josh Groban–soundtracked montage of weeping. But proper, in-real-life crying? Without embarrassment or fear or ridicule? Not so much. Which makes no sense, because sadness is normal and tears are, too. We've always cried. We're *meant* to cry.

"Crying is a way to elicit support from others during times of distress," says Ad Vingerhoets, the "Tear Professor" from Tilburg University in the Netherlands. Humans are the only creatures known to shed emotional tears and infants cry to get the attention of their parents, while an adult might cry to get the sympathy of a friend or loved one.

Scientists used to think that we got rid of "toxins" and stress hormones via tears[19] and that the act of crying also produced endorphins and the feel-good chemical oxytocin.[20] "But there's no change in our pain threshold after crying, which you'd expect with an increase in endorphins and oxytocin,"[21] says Vingerhoets, "and saliva contains stress hormones, too. But who feels better after a good drool?" he asks me.

"No one?" I hazard.

"Exactly!" Vingerhoets and his colleagues found that cortisol levels *do* decrease in those who cry, but that similar effects are observed in baby monkeys separated from their mothers who emit distress calls.[22] So we don't feel better because we're sluicing out toxins: we feel better because *expressing* sadness soothes us. Psychologist Cord Benecke from the University of Kassel in Germany studied criers versus non-criers and found that criers experienced fewer "negative aggressive feelings," like rage and disgust, than people who didn't cry.[23]

"We now know that crying is something all humans are programed to do and that tears serve a purpose," says Vingerhoets. "Charles Darwin famously denied the usefulness of tears, so I see my work as a personal challenge to prove him wrong!" Fair enough.

And—here I hesitate, lest the sisterhood strikes me down—do women really cry more?

"A little," he admits. Testosterone has been proven to inhibit crying, and prolactin—the hormone best known for its role in lactation—also lowers the threshold at which we're moved to tears. "But the messages we get from our peers about crying are also hugely significant," he adds. "We see from research that ten- to thirteen-year-old boys, for example, face a strong pressure not to cry, whereas ten- to thirteen-year-old girls don't. It is more socially accepted for girls to cry." Both men and women have been proven to cry around the same amount over "the big stuff"—like death or divorce: "But women cry slightly more over other things." I press him and he tells me that the core feeling of tears isn't actually sadness, it's "helplessness."

"So we see that women are more likely to weep out of frustration or around conflict because they feel powerless and struggle to express their anger. Even crying when you're scared is to do with helplessness," he tells me. "If you're fearful but you know how to escape the saber-toothed tiger, for example, you can display your fight or flight impulses. But if you're trapped, you're more likely to cry out of a feeling of powerlessness." I tell him this theory doesn't sound great for women. He reminds me that it's not a blast for men, either. "Boys as young as ten are 'learning' that crying isn't acceptable," says Vingerhoets. "So by the time they reach adulthood, it's no wonder many men are reluctant to be seen crying."

There is, however, one domain in which men have traditionally been "allowed" to cry.

From Tiger Woods at the Masters to Michael Jordan crying during his Basketball Hall of Fame speech, sports-related crying has long been deemed "okay."

"There's something almost heroic about crying on the field," says Kees, a professional soccer player friend I quiz about this over wine in my garden one weekend. He tells me about an important game in Prague where his team lost and everyone started crying. "It's sort of accepted to cry over failure in soccer," he explains, "it's allowed. And also, you just do, naturally. You're doing a week's worth of training for ninety minutes on the field. So when it goes wrong"—he clutches his chest—"it's heartbreaking."

I look skeptical at this (I know nothing about sports). "Okay, so you can't compare it to losing friends or family . . ."

No . . . "But it's close."

Despite "heroic" tears, discussions about emotions and mental health have still been frowned upon in sports until recently. "Coaches are now telling us to talk more," says Kees, "that it's important to expose our vulnerabilities so we can bond as a team. People are starting to realize that being more in touch with emotions and being honest will improve the game, as well as the players and soccer in general." A study from Indiana University Bloomington published in the journal *Psychology of Men & Masculinity*[24] found that American soccer players who cried reported higher levels of self-esteem and were less concerned about peer pressure than their non-crying counterparts. And male weepers have some pretty prominent role models, too. Greek heroes Achilles and Odysseus liked a cry. Jesus [*literally*] wept. So we should all cry when we need to.

The first lesson of how to be sad is just to stop fighting it. That's it. That's all we need to do to begin with. Even when it's serious and we still have to get out of bed or care for others: fighting sadness or pretending it isn't there is *not* the answer. We have to *feel* it. Something that sounds deceptively simple but can feel like a radical act since "sadness" is one of the least visible parts of everyday life. We have to get back in touch with our emotions so that we can deal with them.

I know this now. I didn't then.

Back in the 1980s, no one I know wants anything to do with "emotions." So I bury mine.

I throw myself into schoolwork and try with everything in my (albeit limited) powers to be "a good girl" and make my mom "happy," frustrated when my endeavors inevitably fail. I fetch things for her—purses and shoes, mainly. I cling to my mother to show my affection, burying my face in the soft folds of her voluminous 1980s dresses. I bring her breakfast in bed. Once, an entire pound cake. Another morning, a six-pack of muffins. I sit on the end of her bed beaming and watch as she gamely works her way through them. The corners of her mouth rise in recognition of my efforts but the smile never quite reaches her eyes. And then she meets someone who can make her smile. Properly.

2

Lower Expectations

M Y MOTHER HAS a new friend. He has an orange backpack,
plays the guitar and stays the night. I know this because I hear
him strumming Steely Dan after I've gone to bed and see the orange
backpack outside my mother's bedroom door the next morning. I don't
run in and fling my arms around her as usual. Instead, I go downstairs
to make my own breakfast, cereal with heavy cream (to be fair to
child-me, it looks a lot like milk in the carton and I'm new at this cu-
linary stuff). I wait for footsteps on the stairs, which are carpeted, so I
have to listen carefully between slurps of creamy cereal slime. I freeze
when I first see the owner of the orange backpack. He is taller than my
dad. And younger. He doesn't wear a leather jacket! A little like the
way they say that spiders are always more afraid of us than we are of
them, Orange Backpack scuttles off quickly, without making eye contact.
But it happens again the following weekend. And the one after. Finally,
the man with the orange backpack comes for dinner. This is odd since
neither of us has yet committed to making eye contact or been formally
introduced. I want to appear friendly—because "good girl"—but I can't

help feeling miffed that Orange Backpack has somehow succeeded in making my mom happier where I have failed.

I don't mention the new visitor to my dad, but Saturdays and Sunday mornings become my own. I play. I draw. I paint. I craft elaborate structures involving pulley systems, shoeboxes and wicker baskets to transport stuffed animals around the house. And I watch a lot of TV. It is the 1980s so television is relentlessly glossy and aspirational—as are my expectations of life at this point. Okay, so my mom works full time; I'm no stranger to a hand-me-down and "overdraft" is one of the first words I learn. But elsewhere in yuppie-ville suburbia aerobi-cized stay-at-home mothers pick up kids in white Porsches playing "Money Talks"* with maximum bass. Many of my classmates' parents live in huge houses with swimming pools that get upgraded every few years. Face-lifts aren't uncommon (nor are they particularly convincing in the 1980s). Everyone wants to be not just "better," but *the best*. I want a life like that—like the people I see around me and on TV: a bright, shiny, successful one. Unstained. I want a life that is smiley. Happy. *Perfect*. Whatever expectations the people around me hold up, I raise the bar higher still. Because I want the TV-perfect life. Which doesn't seem too much to ask . . . does it?

Orange Backpack eventually moves in, and, for a while, things do become happier. There is music and laughter and trips to London—a decadence that is near dizzying. I am allowed to eat dinner every Sunday in my pajamas in front of the TV, post-bath, letting my wet hair dry by the fire following its weekly wash. Life is lighter. Better, even. Then my dad and his girlfriend announce that they are getting married.

Okay, I think, this is new . . . Be good. Be nice.

I ask if I can be a bridesmaid. On TV, daughters always get to be a

---

* "Dirty Cash (Money Talks)" was a 1989 hit sung by Melody Washington with lyrics that now appear questionable but a banging tune nonetheless.

bridesmaid or, at the very least, a flower girl. I've recently watched a TV show where a girl my age got to wear a tiara and ride a horse as part of her bridesmaid-ing, so I'm aiming high.

But my dad says "no."

I ask if he means the tiara part or the horse part.

"All of it," he says.

I presume he is joking (is this how "jokes" work? Who knows . . .) so decide I will win him over by being extra good and working extra hard at school. Then he'll have to say "yes" because that is what happens on TV, right?

I also get to thinking that if my dad is remarrying, perhaps my mother should, too. There is no question about whether my dad's new wife will resemble anything like a "new mother" (she won't) but I know from television that when a mother remarries, a "new dad" is normally on the cards. I still have an "old dad" of course, but he's not hugely reliable these days so having one in reserve can't hurt, can it? On TV, these blended families all get along famously and live happily ever after. On TV, a mother who has been on her own for a while and looks a little *beige* meets a man with a strong jaw and kind eyes who takes her on picnics and to the zoo, in montage. He wins her over and rekindles something that makes her zhuzh up her act—in the 1980s this generally involves the application of blush, shoulder pads and bigger hair (perhaps a perm). Duly impressed, strong-jaw man then proposes on one knee, at the site of their first picnic, while ducks quack him on encouragingly. Mother and children clap hands in unison with delight and accept. New man swings mother around in an embrace as she points her toes skyward. Stringed instrumental starts up and a wedding montage ensues. The daughter gets to be a bridesmaid, as well as acquiring (probably) a new baby brother and a dog named Bobby. I've seen TV: I know how these things work. Orange Backpack isn't *exactly* like the Ken doll men on TV. But he seems nice and he's what we've got so my mother will, I rationalize, have to marry him. In montage.

I'm hoping this happens sooner rather than later as I come from a long line of Catholics (my great-aunt is a nun) and I already know that adult sleepovers out of wedlock are frowned upon. I now attend a former convent school, which means that as well as mandatory Catholic education, a priest makes occasional spot-checks. When this happens, we hide all evidence of Orange Backpack Man and pretend he doesn't exist. At school, I'm learning all about "original sin," the "sin" of divorce and "living in sin," so I beg my mother to marry Orange Backpack Man, to reduce some of our family's sin-quota and "save us from eternal damnation." But no marriage-montage starts up and no new puppies named Bobby appear. This is a crushing disappointment.

Disappointment is defined as the psychological reaction that occurs when an outcome doesn't match up to our expectations. The greater the discrepancy between the two, the more acute our disappointment. In 2014, Dr. Robb Rutledge from University College London published a study in the journal *Proceedings of the National Academy of Sciences*[1] that included a mathematical equation for happiness based on our expectations. Researchers found that it doesn't matter whether or not things are going well for us—what matters is whether things are going better than *expected*. "It is often said that you will be happier if your expectations are lower," Rutledge said in a statement. "We find that there is some truth to this: Lower expectations make it more likely that an outcome will exceed those expectations and have a positive impact on happiness."

Living up to the expectations of others or living with unrealistic expectations is draining. And having high expectations of other people can lead to disappointment. Feeling "let down" a lot or experiencing a sense of resentment that other people aren't measuring up to the standards we're setting for them is a common sign that we expect too much from them. Of course, sometimes people are just assholes. And we shouldn't allow ourselves to be treated badly by them. But if we have impossibly high standards for ourselves and others—expectations that are consistently unmet—well, then it might just be *us*. And I say this reluctantly, with love and inclusion: it might be *us*.

When we're feeling sad—or struggling *not* to feel sad—many of us would fare better if we learned to expect less of ourselves (and others). But instead, many of us aim high. Really high. And the images of aspirational, glossy lives that we see all around us don't help, either.

Most TV viewers today are a little more sophisticated than I was in the 1980s. We're savvier and we don't believe everything we see. But the rise of Instagram, Facebook and other impossibly well-lit platforms offer many of us another realm where expectations are sky-high. Social media is "a constant bombardment of everyone else's great news, where everyone's showing their best side," says Meik Wiking who conducted research into the impact of social media on happiness—or lack of it. He discovered social media to be a hotbed of performative "happiness" that makes both performer and audience feel worse afterward. "We look at a lot of data and one of the things that often comes up is that comparing ourselves to our peers can increase dissatisfaction," he says. One study showed that participants were 55 percent less stressed after just one week off Facebook.[2]

There are many psychological issues connected to high expectations, including low self-esteem, since failure to meet our expectations confirms our low regard for ourselves. There's also a link to negative core beliefs and the idea that "I have to be perfect to be loved," or "the world is dangerous so I need to be in control," as well as fear of intimacy, since expecting too much of others gives us an excuse to push them away when they inevitably fail to meet our standards. There's also fear of failure—which can lead to self-sabotage—and even fear of change. Some of us fall prey to what psychologists refer to as the "just world" philosophy. "In much of the Western world, we have this idea that you get what you deserve in life," says psychologist Aphrodite Matsakis. "This means that on the flipside, many people think that they must therefore 'deserve' what they get." The idea is that if we're "good enough" or "careful enough" or "competent enough" we can protect ourselves and our family. "Many people have had this reaction but it's important to understand that you are not in charge of this," says Matsakis. "The self-blame is a way to feel you have some control over a negative situation, which might have been totally random."

I struggle with this, growing up. I'm taught at school and at church that I shouldn't mind too much about anyone dying because it means they've "gone to a better place," to be "with the angels" and how: "God must have needed them more than we did." I'm encouraged to believe that sad stuff happens because God "has a plan." Everyone around me seems to get on board with this and continue on unquestioningly, despite what I already perceive to be a world where very little makes sense (fruit in salads?). Life is weird, but everyone else just appears to *cope*. My grandmother nursed sick and dying soldiers in the Second World War; a sick then dying husband in her forties; then "lost" her granddaughter. But she's still going strong and leading a dou-ble-slash-triple life as the Prime Minister Margaret Thatcher and Queen Elizabeth II! My mother watched her dad die when she was a child, then buried her daughter. But now, this Boudicca in a puff-sleeved botanic print looks more radiant than ever! And she has a perm! (Bigger hair: check.) So why aren't I happy? Why can't *I* be happy? Why isn't my life like the lives on TV?

The final, most pernicious side effect of high expectations is perfec-tionism. In 1980s and 1990s middle-class suburbia, perfectionism wasn't seen as a flaw: it was viewed as an advantage—the character trait that showed you were committed. Competitive. A "winner" [*holds hands up to make w-sign*]. Teachers gave praise when you became top of the class. Parents were pleased, too. Being the best at something appeared to come with distinct advantages, I observed. What I gleaned from school, TV, the world around me and parents of classmates whose pools I would swim in was that perfectionism was indeed a worthwhile goal. If not [*whispers it*] the only goal.

Perfectionists, I will later discover, are both born and bred, since perfectionism is a trait influenced by both genetics and our environment.[3] I *might* have tended toward perfectionism without the losses in early life, the religious upbringing, the focus on academia and friends who cruised around in their parents' Porsches. But then again, I might have not. What's certain is that I wasn't an outlier in my proclivity.

Two in five children and adolescents are now perfectionists, according to research into child development at West Virginia University[4] and studies have shown that college-age students today are significantly more likely to have perfectionistic tendencies than students in the 1980s, 1990s or early 2000s.[5] Interestingly, these trends remain when controlling for gender and geographical differences[6]—so it's not a "girl" thing: it's a human thing, exacerbated by our unwillingness to allow for discomfort and failure. And social media only intensifies this. I didn't grow up with social media, but the pressure to succeed was strong. There was a kudos attached to perfectionism and we were all at it. So can it really be so bad?

"Yes," says Harvard University lecturer and author of *The Pursuit of Perfect,*[7] Dr. Tal Ben-Shahar (in the words of "Fame": remember his name. We'll be hanging out with him later). "Of course there are some characteristics of perfection that can still help in terms of success and helping you to be happy," accepts Ben-Shahar, "like hard work, like having a sense of responsibility, like attention to detail. But some of the less desirable outcomes are an innate fear of failure, a lack of appreciation of what we've already accomplished, and rejection of painful emotions."

Perfectionists like to think that the path to success will be a failure-free, straight line, says Ben-Shahar, but the reality is more of a squiggle. As a result, perfectionists are permanently disappointed—becoming harsh critics of both themselves and the world around them. All traits I recognize. Traits I don't like. But herein lies another problem. Perfectionists have a tendency toward low self-esteem. We find faults in ourselves and the world around us, so that we don't much like anything, and then feel guilty and ashamed. It's not exactly a hoot. But more of us than ever are suffering.

The West Virginia University study's lead researcher, Katie Rasmussen, believes that the rise in perfectionism is "heading toward an epidemic and public health issue." Because perfectionist tendencies have been linked to depression,[8] anxiety,[9] anorexia,[10] bulimia,[11] burn-out,[12] OCD (obsessive-compulsive disorder),[13] PTSD (post-traumatic stress disorder),[14] chronic fatigue syndrome,[15] insomnia,[16] indigestion[17] . . . and early death.[18] If this sounds dramatic, that's because it is. Perfectionism, it turns out, is a silent killer.

A better idea is lowering our expectations and swapping perfectionism for something the experts call "adaptive optimalism." This means taking a curvier route through life and enjoying the journey rather than setting our hearts on a poker-straight path and then berating ourselves when we fail to stick to it, says Ben-Shahar. The optimalist still experiences sadness, but this way, we take difficult periods in our stride—because they're part of life and we're reminded that, as the adage goes: "this too shall pass."

The ancient Greeks and Romans had the right idea here and Marcus Aurelius would remind himself every day: "I shall meet with meddling, ungrateful, violent, treacherous, envious, and unsociable people . . . I can neither be injured by any of them . . . nor can I be angry with my kinsman, nor hate him."[19]

In other words: he learned to manage his expectations. The Stoic philosopher Epictetus was even more specific in his advice on how to live, counseling: "When you are going about any action, remind yourself what nature the action is. If you are going to bathe, picture the things that usually happen in the bath: some people splash water, some push, some use abusive language, and others steal. Thus you will more safely go about this action if you say to yourself, 'I will now go bathe, and keep my own mind in a state conformable to nature.'"[20]

There will always be someone who splashes. Chaos is part of life. Scrap that: chaos *is* life. So if we're expecting otherwise, we're bound to be disappointed. As the Danish gloom-king Kierkegaard wrote: *"Life is not a problem to be solved but a reality to be experienced."* But I wanted to "solve" life. To "fix" it. To "win" at it.

Growing up in Thatcher's Britain, lowering expectations was akin to heresy: we aimed *high*. My mother was expected to get on with things, so she did. She started a new relationship; she went back to work; she fed and clothed and cared for her sole remaining child. She even joined the parent teacher association at my school, volunteering for anything going. She made new friends; got ahead at work; got a perm and slapped on a smile daily. And every Halloween, on the anniversary of my sister's death, instead of putting out a pumpkin and welcoming trick or treaters, we would huddle up inside on the couch and watch TV. Every November 1, my mother would clean pieces of shell, yolk and semi-dried albumen from our front door, where we had inevitably been the victims of egging. Incensed by the lack of "treats" and intent on retribution, "tricks" were always egg-based for the children of our neighborhood. It wasn't their fault. They weren't to know that my mom was near suffocating in sorrow: they just thought that she was being churlish, not opening the door to dole out Mars

Bars like the rest of the street. She got to know the neighbors, a little, after a while. But she never told them what had happened. Never mentioned that she couldn't come to the door to greet their offspring in ghoulish costumes because she was still mourning the death of her own child, on this day, all those years ago. She said nothing. We all said nothing. We got on with things. Focused on the positives: like Steely Dan; school and sticker charts. Instead of lowering our expectations, we aimed higher.

This is a trope I see repeating itself time and again into adulthood—both by me and the people around me. After every relationship breakup there is a flurry of activity and self-improvement. I expect too much from myself and others and never get used to the fact that—as Epictetus said—there will always be people who splash at the pool. And cannonball. And heavy pet in the deep end. Or pee in the shallow end.

Two friends going through a divorce in the present day are battling with their own expectations of themselves, just as much as their expectations of their exes. I can see from the outside—from the perspective of someone who cares for them—that this self-flagellation and the blame-game is helping no one. I can't help wishing they'd just cut themselves some slack for a while. Learning to lower expectations sometimes is hard, but it is, according to Ben-Shahar, *important*. And necessary, if we want to be sad, well.

I do not lower my expectations growing up. I do not get to be a bridesmaid, despite working as hard as I can at school and running out of space on my teacher's star chart. I don't get to go to the wedding, either. My dad goes on to have two more daughters and although I'm excited about having half-sisters, this, too, is tricky and makes me feel a little sick again. He lives with his other daughters and sees them all the time and reads to them and does all the things I wish he'd do with me. I want my dad to want to spend time with me and I desperately want him to be proud of me. But seeing him in the flesh every few months starts to be too painful. The knot in my stomach tightens and the disappointment of expectations unmet becomes too much.

My mom comes in while I'm in the bathtub one day—the last time she will ever be permitted in the bathroom during my ablutions. My dad is due to pick me up for one of our sporadic outings and she asks how I feel about it. I stare resolutely at the faucet. She asks whether or not I still want to go and I hear a small voice that doesn't sound like my own telling her that perhaps I do not. And that's it. Arrangements are made. I will only see him again once in the next twenty-seven years.

As far as I know, my father doesn't contest this. And he is seldom spoken of. But I suspect that my mom has had a word with the school this time since a few well-meaning teachers ask about my home life occasionally. One teacher, Miss Foster, tells me that maybe my dad "just wasn't ready to be a father." That "some men never are." I don't have the heart to tell her that my dad seems entirely prepared and enthused by the prospect of parenting my half-sisters. Miss Foster is a kind woman, if jaded. I don't want to see her look sad. Sad, I know, is bad. So I say nothing. Besides, I have things to do. I am impatient for the rest of my life to start and determined that it should start now. Not later: now. I've got to make this worthwhile, somehow. For all of us. I haven't got time for sad.

3

Take Time. Be Kind.

I AM ANXIOUS FOR adulthood to start. I want to *get on*. A day I'm not busy feels like a day wasted and I am horribly conscious of time. Of wanting it to speed up. As Mary Wollstonecraft wrote in a letter to her friend Archibald Hamilton Rowan in 1795: "The slow march of time is felt very painfully. I seem to be counting the ticking of a clock, and there is no clock here."[1]

Or, as I write in my diary in 1992: "Think my watch might be broken. Minutes take aaaages."

I get a weekend job, age twelve, cleaning boats for $3 an hour, start babysitting, then waiting tables, then working in a golf club, picking out the spinal cord from tins of Scottish salmon to make sandwiches for pink-faced men in long socks. This last job I hate so much that I barf in the hedgerows on the walk to work every weekend, without fail. But I'm not about to let regurgitated All-Bran stop me! (It's the 1990s: fiber's very fashionable.) I'm in a race to . . . somewhere!

At school, I sign up for every extracurricular activity going. At home,

I study, read and, once I'm old enough, arrange a frankly hectic social schedule.

"Relax," a friend's mother counsels me one afternoon when I start emptying her dishwasher mid-playdate. "Go outside! Enjoy yourself!"

But I will not be deterred. "Thank you, Mrs. Clarke, but I'll just set the cups on the drainer first." The truth is, I don't *want* to "enjoy" myself or "relax." Why should I? I have a feeling that I don't deserve it. Nor am I entirely sure *how* to relax. Because if I stop, the sad thoughts come flooding back.

There's also a martyrdom to "busyness"—something about the self-sacrifice and the denial of pleasure implicit in always being on the go, always *doing*, that appeals to the Catholic in me. The message I've imbibed since birth is: "Pleasure can come later." With the caveat: "If you've earned it: if you are deserving. If you're *worthy*."

I'm not sure that I am worthy, but—for the most part—at least I'm cheerful. Like Riley, the character in *Inside Out*, I have a strong understanding that my role in life is to be a "happy girl," come what may. Only then comes puberty.

Adolescents are odd and excellent in equal measure, inexperienced in the ways of the world, yet somehow trapped in a nearly adult body that can do things it's never been able to do before. It can feel things and change without warning or any instruction manual.

Aged fourteen, I grow a pair of DD breasts overnight that I am ill equipped to handle, emotionally or physically. Running becomes agony. I have never been sporty but now even hastening my pace to catch a bus becomes impossible. They/I get looked at and commented on. A lot. And because it's the 1990s and none of us know any better than to follow the bro culture streaming out of all available media, we go out and drink. Also a lot.

Aged seventeen, many of us start learning to drive. A few of us are even given control of a vehicle. This is either a parent's, or, if we've saved up enough from dirty boats and salmon spine-removal over the years, an ancient turquoise stick shift, paid for in crumpled bills. My three-door hatchback is a beauty, despite the rust, and I love her with all of my still-functioning heart.

The freedom of being able to leave any situation that I'm no longer enjoying is new and exhilarating. But my impatience reaches new heights. Conversation getting a little deep and uncomfortable? Leave! Bored at a party? Make up an excuse and make it out of there ASAP! For me, driving means liberty. For others, driving means machismo, status, power and peril.

The teenage brain is a work in progress and our adolescence is a time of experimentation and impulsivity, when one impetuous evening can have catastrophic consequences. A few friends take "proper" drugs and one girl in our school dies from an overdose. Three boys from the adjoining school die in car accidents. A girl in my class was going out with one of them. There are red, sore eyes and my classmate is pale. Shaken, even. For days. But within a fortnight she is dating one of his friends. Grief is strange like that. And being a teenager means that we get to feel like the center of our own world. If not everyone else's, too.

Psychologists have found that adolescence involves establishing autonomy from parents or caregivers at the very time our bodies and minds are changing, straining between who we have been as a child and the adult we might become. It's a time of increased vulnerability and hypersensitivity. So it's unsurprising that 50 percent of mental health problems are established by age fourteen and 75 percent by age twenty-four[2]—the age until which we now know the brain continues to mature.[3] That's more than a decade of raging hormones. Thanks, science.

On the cusp of life, there are seminal moments when we stop being a child and this strange new phase of "adulthood" begins. For many these can be fairly benign—part of the natural emergence into maturity. For others, these moments can be traumatic or even tragic. "The talk" is something many of my peers and I got growing up—a checklist on the basics of sex. But years later, I will learn how many Black teenagers around me at this time were getting another, more harrowing version of "the talk" from their parents: that adolescence and freedom also meant danger, that the world was not kind, and that racism was alive and thriving.

I sit next to a man at a dinner party in my thirties who tells me that growing up as a Black teenager, he understood—from his mother and from the injustices he saw all around him—that he was not safe because of the color of his skin. That he would have to take greater precautions, exercise greater restraint and, in many cases, work twice as hard as his peers to get by in a structurally racist society. From then on, he and many others I speak to adopted a form of "double-consciousness," a term coined by the writer W.E.B. Du Bois in his 1903 book *The Souls of Black Folk*.[4] Double-consciousness describes the inner conflict and duality required to live as a person of color in a predominantly white culture: looking through one's own eyes *and* the eyes of the people in power. Learning early on to view oneself how white people might and so walking differently. Talking differently. Dressing differently so as not to appear "threatening" and play into prejudices. Or get into trouble with a teacher. Or the police.

Quite apart from the horrific injustice of this, living with a duality of consciousness comes at a cost, disrupting the idea of who we are and how we value ourselves. Cognitive dissonance occurs when we hold two or more contradictory beliefs, values or ideas about ourselves—and this results in psychological stress. The mental load and daily effort of double-consciousness is finally being expressed in popular culture and talked about by writers such as Reni Eddo-Lodge,[5] Afua Hirsch,[6] Akala,[7] and Yomi Adegoke and Elizabeth Uviebinené[8] (more from Adegoke in Chapter 6).

Jade Sullivan is a Black activist, writer and entrepreneur I speak to after reading a piece she wrote on racism, the Black Lives Matter movement and growing up mixed heritage.[9] Her mother is white and her dad is Jamaican, Sullivan tells me, "so I was always aware of being mixed heritage. I identify as Black as this is my culture and the way the world sees me. I remember one of my earliest childhood memories was walking with my mother and saying to her, 'Mummy, why can't I have blue eyes and blonde hair and look like you?' She was always open and a teacher and a lover of Black culture, so she told me, 'One day everyone will look like you.'" (She was right: the mixed-heritage population is the fastest-growing ethnic group.) "'Why do you think white people sit in the sun to get a tan and perm their hair?' She told me I was beautiful." But the rest of the world wasn't always kind. Aged eight, Sullivan went to a swimming pool one day and remembers someone calling her what Sandra E. Garcia describes in the *New York Times* as "the unutterable slur that rhymes with bigger."[10] "Then there was my best friend from the age of five," Sullivan says. "She had curly red hair, freckles, and looked just like Annie. We were constantly at each other's houses. We'd have sleepovers. We'd spend our weekends together. Then when we were going to high school her parents said, 'You can't be friends with Jade anymore.' They worried that my friend would end up going out with Black men if she hung around with me," says Sullivan. "Fortunately," she adds, "my friend ignored her parents—we're still best friends. But there were daily encounters with racism growing up. It takes its toll. Life gets serious pretty quick when you're Black. It has to. Unfortunately they [Black people] don't have the luxury of skipping through life being ditzy." I take this on board. It weighs heavily. Sullivan concludes: "I don't think you can actually quantify the effects of daily racism on Black people's mental health."

Mitch Prinstein from the University of North Carolina believes that our experiences in adolescence have a profound effect not only on how things go for us growing up, but on our *entire lives*. In *Popular: The Power of Likability in a Status-Obsessed World*,[11] he writes that our experiences during our teenage years change "our brain wiring" and, consequently, "what we see, what we think and how we act."

For me, school is okay.

I acknowledge all my privilege here. Being a geek does not exclude me from "the in crowd." There are only twelve pupils in my year, so it's tricky to pull off any kind of "crowd" (in fact, the school is so small, it will soon face closure). For the most part, we all tolerate each other. In addition to this, the "overnight boobs" mean I do okay in terms of "teenage girl status." But a girl at the neighboring school finds the arrival of puberty more problematic. When she grows breasts, she has "slut" written on her locker in permanent pen—as though she has deliberately grown them for the sole purpose of arousing lust in others and should be "shamed" for doing so. She rubs at it with nail polish remover pads but the scrawler strikes again by morning, engraving insults with a compass this time. My friend Dave has his head flushed down the toilet by bullies at school. Every. Single. Day. He develops the reflexes of a gazelle, able to pull off his glasses *just* before a hand pushes his head toward the bowl and the handle is yanked. This way, he tells me, he avoids the reprimands of his father for having broken another pair of glasses, on top of the indignity of wet toilet-water hair. The bullies were never punished and never seemed to realize the hurt that they were causing.

Most of us know anecdotally—even intuitively—that the effects of intimidation, harassment, mistreatment and discrimination last for many years. But I will later learn that bullying and depression go hand in hand and that children who are verbally and physically bullied are at far greater risk of developing mental health problems.[12] Studies show that the experience of being bullied, rather than being left behind on leaving school, may linger and affect the victim's life well into middle age.[13] Adolescents who are involved in bullying—as a victim, bully or both—are far more likely to report feelings of low self-worth and sadness and to feel unsafe in their school.[14] Experiencing racial discrimination has long been linked to poor mental health outcomes and there is a large and growing body of evidence demonstrating that racism leads to mental illnesses, especially depression, prolonged grief or difficulty coping with and adapting to severe events.[15] Oh, and trying to "get over" racism by ignoring it only makes things worse. Just as the Wegner research showed how trying not to think about "sad" only increased the "sad," there's research to suggest that trying not to think about race as a person of color may increase stress. A 2014 study published in the *American Journal of Public Health* investigated the impact of reported racism on the mental health of African Americans at cross-sectional time points and longitudinally, over the course of one year and found that: "individuals who denied thinking about their race fared worst."[16] Injustice sticks. We can't ignore it—and the hurt hangs around for far longer than we might expect.

The girl who has her locker defaced never recovers her self-esteem. The boy in the glasses will go on to feel a need to prove himself, at every turn, for the rest of his life. Never quite feeling "enough." "Every day it's like I'm trying to show the bullies that I can make something of my life," he says, "to prove that I'm not worthless." It doesn't matter how many times friends and loved ones assure him that he is not worthless—that no one is—it's how he feels. And it's heartrending.

There is no internet when I'm growing up and I pass through youth treated pretty well by my peers. But I don't shake off the high expectations on myself and others, the proclivity for "busyness" or the idea that I have to earn my place in this world. During my teenage years, I also add some pretty unwoke metrics of feminine success to the standards I set for myself. I want boys to like me and I'm as predictably eager to win male approval as any "child who's missing a father cliché" you might hope to stumble across. I have shrewdly been placed in an all girls' school, but once I'm deemed old enough, I'm allowed to get the bus to school WITH BOYS. I see and sometimes even *speak* to boys from the adjoining school between the commuting hours of 8 a.m. to 9 a.m. and then again between 3 p.m. and 4:30 p.m. There are also boys in a few of the extracurricular activities that I have predictably signed up for. And when I say "a few," I mean "all." Academic success stops being my everything: men matter now, too. I find a diary from 1997, aged seventeen, where I summarized the highlights of each day, as follows:

> *Thurse* (sic), *13 February 1997*
> *Mr Gildersleeve said I was thin.*

Oh dear . . .

> *Wednesday, 19 February 1997*
> *Richard Toombes remembered me.*

> *Tuesday 25*
> *Lee asked me out on a date.*

> *Thurse* (sic) *27*
> *Lee asked me out on a date.*

Gosh, I think, Lee's eager. I mention this to an old school friend in the present day and she reminds me that Lee was in fact a thirty-two-year-old man who worked with a friend's dad. And he was asking out a seventeen-year-old. Urgh. Go home, Lee.

The diary continues:

Thurse (sic) *6 March*
Simon spoke to me.

Simon, at least, is my age. Though for a studious girl I suck at spelling.

Sat 8 March
My driving instructor said I wasn't too bad and had nice hair.

This is Kevin, the driving instructor who also puts his hand on my knee (#metoo) and tells me he doesn't think girls should go to college because they're "only going to stay home and have babies." I essentially pay Kevin $14 an hour to compliment my hair and administer weapons-grade sexism.

Thursday 20 March
Benji offered me a ride home.
I wrote a play and a playwright man said it was very good.

Wait, what? A "play"? *That was "good"?*

I have no recollection of this. And yet Kevin I remember? Thanks, brain . . . If I did indeed write an *actual* play that was read and appreciated by an *actual* playwright, I think it should probably have higher billing on the day's bulletin than as an addendum to a ride home with a boy called Benji.

At this point, the page has blurred patches on it where the ink has started to wash away, as though water has splashed onto it. Or tears. All I remember from this period, aside from Kevin, is a tugging dissatisfaction spurring me onward and an urgency to grow up. I want to achieve things and make boys like me. I feel guilty for being alive when my sister is not, and so I push myself harder and harder.

Guilt over "not dying" when someone else in our lives *has* is surprisingly common, says Dr. Hannah Murray, research clinical psychologist at the Oxford Centre for Anxiety Disorders and Trauma. This seems unfortunate since by dint of reading this now, we are alive while others we've known and loved are not. "Survivor guilt" as it's known has been around as a medical concept since the 1960s. "There were high rates of PTSD and survivor guilt in Vietnam veterans," says Murray. "Survivor guilt can often lead to self-harm, self-sabotage, or a feeling that we must almost 'repay' a debt." And it doesn't have to be linked to death. "One definition of survivor guilt is feeling that you have an unfair advantage over someone else," says Murray, "so it could be that you survived a mass redundancy, or that you've simply had opportunities others haven't. There have been studies about first-generation college students experiencing survivor guilt and it's also been documented among gay men who tested HIV negative in the 1980s." Murray is currently preparing for what is likely to be a wave of PTSD and survivor guilt in patients, staff and family members who have lost someone to coronavirus. "Especially if they weren't able to be with the person who died to say goodbye," says Murray.

At one time, survivor guilt was included in the American Psychiatric Association's *Diagnostic and Statistical Manual of Mental Disorders* (remember this?) although in the most recent *DSM*, survivor guilt isn't mentioned at all. But many more of us than we might expect will suffer from survivor guilt at some point—whether or not we've suffered a trauma, says Murray. We may feel ungrateful for the life we have, thinking, I have all this, why should I complain about feeling sad? But we all get to feel sad. This is normal and we have to acknowledge and accept these feelings rather than trying to prove ourselves all the time.

I have an acute case of survivor guilt. For "me" to be worth it, I feel as though I have to work twice as hard. My addiction to over-achieving continues and my impatience reaches new heights.

Adolescence is a time when many who've experienced a sad event in childhood can find things tough. Because here's the thing: there's no time frame on grieving. And, apparently, keeping busy to the exclusion of "feeling" isn't the best plan.

"It's not an unusual response," says Julia Samuel, "this urge to 'achieve' and 'keep going.' It's fine to be busy—as long as you're not busy to avoid facing up to something else. Especially if that something is sadness. If you're cramming your days with activities and appointments and deadlines to distract yourself from feeling and experiencing your emotions, you'll pay for it in the long run."

We need to allow time, be kind and practice humanity—toward ourselves and others. We need to give our sadness time to breathe.

"If we don't address sadness we store up problems," says Ross Cormack from Winston's Wish. "There's lots of evidence it can have a lifelong impact, even leading to depression," he says. "And the longer the feelings are stored up, the harder it is to unpack that. Neuroscience shows that if we go through traumatic life events that we are not supported in, it can impact how a person's brain grows."[17] This is especially true of grief.

In the present day, a friend's mother has just died. He is currently consumed by sadness and loss, in a place where there are no answers and nothing seems to make any sense. But I can already see that he's urging himself to be "done" with this and move on. He's journaling, scheduling, planning. He's eager to "move on" to the "next phase" as he calls it. But this is a myth. Many of us will be familiar with the Five Stages of Grief or the Kübler-Ross Model, first introduced by the psychiatrist Elisabeth Kübler-Ross in her 1969 book, *On Death and Dying*.[18] Kübler-Ross wrote that there were five phases: denial, anger, bargaining, depression and acceptance. Although regularly cited in popular culture as the necessary steps for how to grieve a loved one, the stages were actually intended to describe the series of emotions experienced by someone who is dying. They're for the soon-to-be-mourned, not the mourner. The stages have never been empirically proven and are now largely viewed as outdated.[19] Later in life, Kübler-Ross herself regretted having written them in a way that was misunderstood,[20] accepting that grief isn't so neat as to be conveniently tied up into stages. When we experience a loss, there's a very good chance that it will never be entirely all right. It's painful and it's hard, but we may never get that cinematic closure that Hollywood movies promise us. However much we want it and however much we try to "speed up" our sadness. We have to take our time, rather than "filling it."

"We don't allow time for grief in our society. We approve if the bereaved person is 'brave' and just gets on with things—and disapprove if they don't. But grieving takes longer than anyone wants or expects," says Samuel. "We can't fight it, we can only find ways to support ourselves in it. When we block it, there are much higher rates of both physical and mental illness. On the positive side, over time, the intensity of the pain lessens, and we do naturally adjust and re-engage with life again. But there may be times, decades later, when an anniversary, a sight, a smell, or a new loss, triggers our grief and makes it feel as new as the terrible day the person died." Cormack describes this as "growing around" your grief, rather than "growing out of it" or getting over it.

Sadness doesn't go away. But we cope better if we allow time for it, rather than "busying" ourselves continually. We are not the sum of what we do: we are not worth any less because we are not achieving all the time. Sometimes, we just need to *be*.

For this book, I speak to the British broadcaster and journalist Jeremy Vine, a man I have always found funny and frank. He has spoken out about experiencing what he describes as a "circuit breaker" of a year, during which time he felt "abject misery" and sought professional help. I'm interested in finding out what he's learned—from his own experiences and from the daily phone-in show he hosts on BBC Radio 2. He's an extremely busy man, hosting two shows back-to-back every day, one on the radio, one on TV, as well as a long-running quiz and covering elections. He's also a fifty-four-year-old father of two and a husband, and a son who recently lost his dad.

I want to know how he handles feeling sad and the tough periods life throws at us. He tells me: "I have to allow and build in time for these." I assume he means by engaging in practices like meditation or going off on fancy retreats. But no, he tells me: what he means is that he actually *anticipates* periods of sadness now. "About once every five years I'll have a massive fuck-up. I know that will happen," he says, "so at the start of the year, I'll say I've got one year ahead of me: I've got two kids. I've got X number of days I have to work for *this* organization; X number for *this* organization. And I'll build in the cock-up. You know—like a death of a parent. Then I'll leave some time for myself." By allowing for these feelings and allowing the time he needs to feel them—planning for them, journaling them, even—he is better able to cope.

"You have to accept what you're feeling," agrees Samuel, "and be patient. Dealing with trauma and grief is something that takes time, and the process isn't always linear." This isn't what I want to hear now and it's certainly not what I'd have thanked anyone for telling me back when I was a teenager. Because who aspires to be "patient"? "Patient" isn't cool. And it isn't valued much in modern life.

Philosophers and religious figures have long praised the virtues of patience[21] and scientists have found that patient people really are better off than the rest of us. They're more satisfied with their lives,[22] by and large, and experience lower rates of depression.[23] This could be because they can cope with upsetting or stressful situations and are good at hanging on during times of hardship. Patience makes us more hopeful, resilient,[24] cooperative, empathetic, grateful, forgiving[25] and generous,[26] too.[27]

Blessed are the patient, for they can . . .

. . . *wait*.

Patience is also associated with popularity and staving off loneliness, because making and keeping friends—with all their weirdness and propensity to tell the same story again and again—requires tolerance and a degree of restraint.[28]

Evolutionary theorists think that we evolved to be patient because tolerant, tireless cave dwellers outlived their more impetuous prehistoric colleagues—and learning to wait is conducive to cooperation rather than conflict.[29] While our impatient ancestors butchered each other before they had the chance to pass on their genes, the ones who were lovers not fighters, had all the sex, all the babies, thrived—producing the descendants we know and love (maybe) today. So if we want to survive and thrive as human beings, we should be able to "do" patience, too.

But patience levels have plummeted in recent years thanks to the pace of modern life, same-day delivery, internet speed and social media. Ninety-six percent of Americans are apparently so impatient that they knowingly consume hot food or beverages that burn their mouths, according to a 2015 survey by the Fifth Third banking group.[30]

I am not naturally patient (can you tell?). I don't like sitting still and I hate flights, long train journeys and movies of more than ninety minutes (Peter Jackson: why?). Fortunately, we can learn to be more patient by training ourselves. Sarah Schnitker invited undergraduates to participate in two weeks of patience training,[31] where they learned to identify feelings and triggers, regulate their emotions, empathize with others and meditate (there's a lot of science saying it's a good idea so if it works for you—go forth and Ommm). In two weeks, participants reported feeling more patient toward the ~~annoying~~ "challenging" people in their lives as well as less depressed. So patience is a skill we can practice.

The Harvard University art historian Jennifer Roberts[32] believes "immersive attention" is such a key skill that she makes all her students choose a work of art, then go and look at it—for three hours. She accepts that while this may feel like "a painfully long time," it's a means to help overcome the initial jumpiness of inaction, learn to tolerate it and come out stronger. She believes that the "deliberate engagement of delay" is something we should all learn in modern life. Because patience is mastery of our own discomfort—and it's a superpower.

Which is good to know—*now*.

Rewind to the 1990s and I am so scared of even attempting to tackle unpleasant sensations that I refuse to acknowledge them. Impatience is my modus operandi and I become a moving target, keeping fast and low. I am nimble, never sticking with anything long enough for unnecessary attachments to develop or tie me down. I do not allow time: I try to *beat* time. And I appear to be rewarded for this impatience, both financially and in terms of approval from my teachers/family/society.

Then everyone dies.

This is an exaggeration (as befitting teenage reflections). In truth, *three* people die (but you know, "three"/"everyone"—who's counting?). I wouldn't want you to think that we were permanently burying people in my family, but unless you've been incredibly fortunate, it's likely you'll have done the same over the decades.

My paternal grandmother dies first. It is decided that I should not go to the funeral but merely "mourn from afar," whatever that means. Then my maternal grandmother who looks like the Queen/Margaret Thatcher dies. Since she is one of the two central figures in my life, along with my mother, this is devastating. Her funeral is long and Catholic and "very moving." I do not cry. Next, to complete the hat trick, my paternal grandfather dies. A portly man of the old-school mold, who drank beer from a huge mug, didn't think a meal was a meal unless it contained meat and was adored by all who knew him. Including my mom and me.

My mom takes the day off work and drives me to the funeral, three hours away. I'm not scared about going to funerals—I'm already less afraid of dying than I am of living. But I haven't seen my dad for years at this point and I feel nauseous at the prospect (nothing new there).

Because I'm on a constant quest for male approval and because I think that there might be boys at the funeral, I select an ill-advised outfit of knee-high boots, a pencil skirt with a slash to the thigh and a stretch pinstriped shirt that strains at the bosom. It's quite the ensemble, though, in my defense, it's the 1990s and the pinstripe is intended as a hint to my "maturity."

The funeral is "very moving." I do not cry. There is mingling in the parking lot afterward and I give my dad a nervous wave.

He gives me a nervous wave back. Some long-lost cousins and my half-sisters do shy waving, too. Soon, we are all waving, in mute.

"Hi," I manage, finally, from beneath a waterfall of hair.

"Hi," he responds.

"Hi," cousins and half-sisters rejoin.

It's not quite the "running into each other's arms and being swung around to the sound of violins and something rousing from Hans Zimmer" scene that I had imagined in my mental storyboard of this particular reunion. A few of my dad's relatives who haven't seen my mom in fifteen years but remember her fondly get chatting. We are invited back to my grandfather's house for the wake and are just preparing to get into the car and drive there, when my dad takes my mom's arm in a gesture of alarming intimacy.

What's happening? Is he going to declare his love for her? That it's all been a terrible mistake? That we can stop lying to the priest about Orange Backpack Man?!

"Everything okay?" My mom also looks startled. "We'll see you back at the house."

There is some mumbling. My dad is saying something.

"What?" My mother speaks sharply.

"I said, I'd rather you didn't," he repeats, louder this time. "Come back to the house, I mean." He explains that he thinks it would be better for all concerned—that it would make everyone feel "more comfortable"—if my mother and I left: "now."

I will later appreciate that a man, mourning his father, may do and say some strange things. I will later understand that we all go a little off balance in times of grief. Or when faced with both women we've promised to stick with "until death us do part" in the same cemetery parking lot. Or when faced with a daughter who's been largely forgotten about in the presence of two newer, shinier "resident" daughters. Any number of thoughts could have been galloping through his grief-addled mind, I realize now. But I'm afraid I don't make any allowances for this back in 1998. All I think is this: that the father who rejected me time and again, who hasn't seen me in five years, is now begrudging me a sausage roll at my grandfather's wake.

Everyone's dead and my dad doesn't want me.

What happens next happens in Instagram slumber filter, at least in my memory. I do not react with humanity or kindness—or allow time to process the tsunami of emotions I'm currently experiencing. Instead, I call my father something beginning with "c" and rhyming with "hunt." Loudly. In front of his two other daughters, wife and a crowd of mourners—many of whom are also my relatives. I am encouraged into the car by my mother and we speed out of the parking lot, spraying gravel in our wake. I am not proud.* I am certainly not happy. My

---

* I would like it noted for the record that this is the one and only time I've used that word as a term of abuse, as opposed to a rousingly robust gynecological descriptor.

mom and I pass the three-hour trip home in silence. We won't speak of this again for years.

The following week I chop off my waist-length hair to a sharp bob, do something unspeakable to it with Sun In,* acquire an unsuitable boyfriend and change my surname by statutory declaration.

If my dad doesn't want me, I don't want his name anymore, is my rationale.

With a new name, new (horrible) hair and a rancor-fueled rocket up my ass, propelling me forward, I am launched, hurtling into adulthood. I am my own special creation. I am George Hearn in *La Cage aux Folles*, only with Sun-Inned hair and cargo pants. I am Sandy in *Grease* . . . Only with Sun-Inned hair and cargo pants. I am . . . reborn! I am "fixed." Right? RIGHT! Like a tethered horse, straining to get free, I tense my muscles to bolt. Rather than hanging around being sad or even contemplating any of the reasons I *might* be sad, I rush into adulthood at full speed. I'm on fire! I'm ready! I am not ready.

---

* For the uninitiated (lucky you) Sun In is an over-the-counter bleach-based spray that was very popular in the 1990s. Just spray onto damp hair, sit in the sun for a few hours or pop on a hairdryer and *ta da*! You too can have hair like straw . . .

4

Avoid Deprivation

I THINK I WANT to be a journalist. No one I know is a journalist. Or knows how to be a journalist. Or reads newspapers. So I've got some work ahead of me. But I'm naturally curious (read "nosy") and an aptitude test in our sole "careers" session at school suggested that I *might* like to consider a career in the law or journalism. I've watched plenty of legal dramas and although there seems to be a lot of Scotch and sex involved, being a lawyer seems to involve more self-doubt than even I can handle. This leaves journalism.

I buy some newspapers and write to anyone with a friendly looking byline picture to beg for work experience. I'm aware that this is half-baked, but if you don't know any better, it's the equivalent of asking someone for directions in the street: you go for the ones who look the most approachable. The ploy works. Ish. But I'm already signed up for a humanities degree when various Actual Real Life journalists I encounter tell me that I really should do "vocational training" to enter their hallowed profession—and my degree is worth nothing.

I have a lot of fun at university, making the best friends a girl could hope for and accidentally dating a surfer with a six-pack you could strum like a guitar (#goodtimes). But the cost of tuition fees and accommodation means that I am, as my friend Ian puts it, "balls-deep" in student loan debt by the time I graduate and am spat out into adulthood.

I can't afford "vocational training" straight away. I can't afford a vessel to urinate in. I can't study more without working first. So I work and save up for two years before starting a journalism postgrad program. I'm also working behind a bar and in a movie theater to pay my London rent. This doesn't leave much money for food so I subsist on popcorn and half-eaten family-size bags of candy that "customers" leave behind.

By now, many of my peers are lawyers or teachers or working in finance. One's even having a baby. Whereas I am a student—again. My status could scarcely be lower. I have a sneaking suspicion that the glittering future teachers and family expected (or rather, that I expected of myself) hasn't quite come to fruition. I worry it never will.

Added to this, my mom has recently split up with Orange Backpack Man. She is so upset that I have done what every kind and caring daughter would do: I have worked very hard to ignore what's happened and get on with MY LIFE. Mostly because I can't bear the alternative: I don't know how to engage with her pain without opening up myself to it, too. If I open myself up to feeling sad, I have no idea what will happen. I say I have no idea, but I'm *pretty* sure it will involve lightning, raining frogs and general Armageddon.

Luckily for me, I am *fine. FINE!*

I am an adult now with a hot, supportive boyfriend of my own who I love and who loves me. He's my first love, the boy I started dating after my grandfather's funeral (that's right: the unsuitable one!). We split up while we were both at college but now we're back together—I presume "forever." I have swallowed the fairy-tale princess manual whole and this is how my story goes, I feel sure of it. Okay, so he hasn't been so supportive recently: he canceled on me last night to play poker. And he seems rather noncommittal about the idea of spending Boxing Day together. But, well . . .

I have a hot boyfriend who I love and who loves me!

Only, I'm not sure he loves me all that much. And he's developed some rather lax approaches to personal hygiene of late. Still . . .

I have a boyfriend! So that's something. *Isn't it?*

I am trying to write an essay in the college library in a poor attempt to distract myself from the fact that my boyfriend hasn't responded to the last message I sent him, but the internet appears to be out. The idea of an IT malfunction is far more appealing than the alternative: that he is ignoring me.

I haven't heard from my ~~hot, supportive~~ boyfriend ~~who I love and who loves me~~ for two days now. I refresh my Hotmail a third time, and, ha! There it is! An email from him. I click to open and immediately wish I hadn't.

The internet, it seems, is working: what's not is our relationship.

"I just don't think this is working out . . ."

I feel a punch in my gut. Am I being dumped? By *email*?

"It's not you, it's me . . ."

Oh come on! That old chestnut? Really?

"You're 90 percent perfect . . ."

Wait, what?

". . . but it's not enough for me."

Whumph.

I feel as though my internal organs have been drop-kicked. But there it is. On screen. My basic, overly neat but impossible-to-shake childlike fear is confirmed: if I'm not "perfect," I will not be loved. I am not worthy. I am not *enough*. Sooner or later, everyone will leave me. I had suspected as much—and the prophecy has been born out.

What now?

I feel ephemeral and loose-limbed and sick and out of control as I get up and make my way out of the library. A storm is brewing outside, tumbling dirty pigeons and dark clouds together. Rain plummets. There is no sun. I float back into the student cafeteria in something of a daze to join the rest of my classmates.

One of them is licking the inside of a bag of potato chips while another tears open one of the free packets of sugar, for lunch. We are all poor. We are all supposedly "mature" students, aware we're being outpaced and outearned by our peers. Sugar packet woman explains she is eating this both for economic and aesthetic reasons: she's broke *and* she's determined not to gain as much weight as she did when she was an undergraduate (our course doesn't cover "nutrition"). Chip-licking man has rationalized that if he can't afford nice clothes for a while, he may as well just stay in shape so that at least the clothes he does have look good on him. I feel as though I have lost what little hold I had over my life, but it occurs to me now that what I choose to put in my mouth is something I can control.

If I can't be the best and win stuff, I could at least, perhaps, be the thinnest . . .

With this twisted logic as my lodestar, I stop eating leftover candy and popcorn. Over the next few weeks I refine my approach to something just shy of "fasting" but far below "a healthy intake of food." Needless to say, I am hungry. All the time. But it's as though I'm finally suffering and "paying" for whatever it is I feel I'm guilty of. Self-deprivation and denial suits my Catholic sensibilities and I become very "good" at not eating. I create rules for myself about where and when I can eat and strict punishments when I (predictably) fail to live up to these strange self-imposed "standards." Now, I can add the new emotion of disgust to my catalog of shame. To this day, I can recognize the anxious, flickering look of a fellow sufferer from disordered eating. Quite aside from the face shape and downy skin, there is a wide-eyed, wary look to someone who has a problem with food.

My breasts diminish. As someone who has had to wear a sports bra even to trot for a bus since the age of fourteen, I find I quite like the sensation of being lighter. I am no longer pneumatic: I am streamlined.

Because I'm now lacking subcutaneous fat for insulation, I am constantly cold and get sick a lot. I also get very little sleep. This is partly due to the ridiculous number of hours I'm working and partly due to the fact that when I try to sleep, I can't. It's my first experience of insomnia but it won't be the last.

One afternoon, against the roar of rush-hour traffic, I collapse outside a bar near the flat I'm renting a room in. I struggle to breathe and become convinced that I am dying. Possibly of a heart attack. I spend several minutes trying to remember how to breathe, before a pair of sensible shoes materializes. A woman bends down and asks if I'm all right. I am not all right. In fact, I'm all wrong. She seems to guess as much because instead of waiting for a response, she hoists me up then fishes for something in her bag. In spite of the sensible shoes, I'm slightly worried she's going to pull out a weapon of some description. Strangely, however, this doesn't seem to bother me too much.

Oh well, I think, would it be so bad? I am very tired. Goodbye, world . . .

But the woman's hand emerges with nothing more threatening than a Werther's Original. I've been warned never to accept candy from strangers but, then again, I've been playing by the rules all my life up until this point and they don't seem to have done me any favors. So I take it and suck, until I can feel my feet again.

After a few moments, the woman asks me if I'm okay. "I'm fine!" I lie, embarrassed. I promise that I'll look after myself. That I'll "take care." But she can't know what she's really asking: that I'll "take care" of twenty-two years' worth of sad that has just come crashing down on me, demanding attention after decades of neglect. I wouldn't know where to start. Besides, I've got a job interview later this week for a writer position. It isn't for the publication I'd been hoping for, but it's a start and it's money. I need this job.

I make it home and sleep fitfully but by morning I've got tonsils the size of golf balls (tonsils are my kryptonite) and a temperature of 101. So I call my doctor.

The doctor checks my throat and confirms that I'm going to need antibiotics.

Then he picks up a syringe and announces that he wants to take blood.

Unusual, I think. But I am not a medical professional and so allow a tourniquet to be tightened above my elbow and try not to look as a needle goes in, very slowly.

"Tell me about your eating habits," he asks next.

My eating habits? *Maybe I'm allergic to something*, I think. *Maybe that's why I keep getting ill*, I think louder. Everyone has an allergy these days, don't they? Maybe, if I just lay off the lactose or whatever, I'll be happy and successful and no one will leave me, ever again!

I do not find out about an allergy. Instead, I am ushered onto the scale.

What? Why? This isn't what I'm here for!

"And how you have been feeling?"

"Fine!" Well, apart from the not sleeping. And the panic attack . . .

"And physically?"

"Fine!" I insist. Apart from being cold all the time. And bruising easily. And having to have a pillow between my knees at night in case I roll onto my side and my bones clink uncomfortably together . . .

He asks about my thoughts, feelings and behaviors around food.

"Fine . . .?" I mumble.

For an aspiring journalist I have suddenly developed a worryingly limited vocabulary.

"And are you regular?"

I tell him he could set his watch by my bowels. This, apparently, isn't what he means.

"Are you menstruating?"

"Oh! That!" Now I come to think about it, not in a while . . . I had thought that this was a bonus (think of the money saved on tampons each month!) but the doctor gives me a look that suggests perhaps it's not.

"This is for the antibiotics." He hands me a slip of paper, then frowns and hammers something into his beige keyboard. "And I think we should try to get you back up to [*insert low number here*] pounds as a priority."

"Oh!" *Ohhh . . .*

My first instinct is to tell you how much I weigh at this point as well as details of the injudicious diet that has led me here. But having worked on this chapter with an eating disorder charity, I hope you'll understand why I've decided not to. If you don't have a problem, chances are you'd find my daily diet and vital statistics at best "weird" and at worst "dull." If you do have a problem with eating, the specifics of another sufferer's diet or weight can pose a risk or be used as a goal. That's how it works. That's how distorted our approach to food becomes.

I haven't got a scale at home. I haven't got a full-length mirror either. I have no way of assessing how thin I have become at this point other than the clothes I can now pull on and the fact that I've had to start wearing belts. With extra notches. I have no clue that things had got so bad.

But I also experience an unnerving glimmer of . . . what is it? *Pride?* Have I "won" at "thin"?

I haven't won anything: I have anorexia.

And I am dismally textbook.

"Competitiveness, perfectionism, seeking control and low self-esteem form some of the key personality traits that raise the risk for eating disorders," Tom Quinn from the eating disorder charity Beat will tell me years later. In a *Clinical Psychology Review* article on the link between eating disorders and perfectionism, psychologist Anna Bardone-Cone and her colleagues cite research suggesting that "the aspect of perfectionism associated with the tendency to interpret mistakes as failures is most strongly associated with eating disorders."[1] Perfectionists are particularly susceptible to eating disorders because in the grips of an "all or nothing mindset," only extreme failure or extreme success exists. So if the perfectionist is concerned about body image, the choice that they see for themselves is between bingeing or starving. There is no healthy middle ground.

Beat now acknowledges seven types of disordered eating, including avoidant restrictive food intake disorder (ARFID), binge eating disorder (BED), bulimia and the increasingly common "emotional eating."

"It's normal for people who aren't suffering from an eating disorder to choose to eat a bit more or 'overindulge' sometimes," says Quinn. And the use of food as a comfort is universal. I learn in Tiffany Watt Smith's *The Book of Human Emotions*[2] that the Baining people of Papua New Guinea use the same word for hunger and the fear of being abandoned (*anaingi* or *aisicki*), so close is the connection between physical hunger and the desire to be cared for. The Germans even have a word for the weight gained from emotional eating—*Kummerspeck*, which translates as "grief bacon." "But if it becomes a pattern or something that happens regularly, it's a problem," says Quinn. Whereas physical hunger comes on gradually and is satisfied when we're full, emotional eating comes on suddenly, feels like it needs to be sated instantly, leads to cravings, isn't satisfied with a full stomach and causes feelings of guilt, shame or powerlessness. "In other words, it's emotional and it doesn't make you feel good," says Quinn.

There's also orthorexia, defined in 1997 by Colorado-based occupational medicine specialist Dr. Steven Bratman as an unhealthy obsession with eating "pure" or "clean" food. Although not currently recognized in a clinical setting as a separate eating disorder, orthorexia is becoming more and more well-known (take the official Bratman Orthorexia Self-Test in the Notes at the end of the book if this sounds familiar).[3]

"Eating disorders are serious mental illnesses that can have lifelong consequences, or even be fatal," says Quinn, "so treatment should be targeted and swift." Fortunately for me, it is. I am fast-tracked to see a cognitive behavioral therapist who is also a qualified dietitian within two weeks.

My therapist is a lovely lady with a bowl haircut who listens, makes a few notes, then gives me a strict eating regime to get back up to "target weight." It feels overwhelming but I am nothing if not obedient. I am instructed to eat, so I do. The day of my first "session," I also get my first job in journalism. The phone call offering me a writer's job on a weekly magazine means that I can now quit my bar job. And my cinema job. Taking up full-time employment on a decent wage means that I can no longer blame poverty for my purging: from now on, it's just . . . me. So I eat. But I still feel shame for "failing" at my warped goal of "not eating." And I feel shame for actually enjoying food again, as though I don't deserve to. I eat in secret; in a way that looking back now seems absurd and unfathomable—just as it must to anyone who's never had a problem with food. But at the time it seems wholly rational. I eat cake in restrooms. I eat a whole can of baked beans on the subway. I eat my roommates' cereal at 11 p.m. (I am very lucky indeed that some of these women remain my best friends). I pass the doctor's weight target [*sounds trumpet*] but the therapist would like to see my scale go up a little higher. She wants me to keep eating. So naturally I obey but then spend a large sum of my newly acquired salary on a gym membership. I feel as though I have "gamed" the system: sure, I'm eating more but I'm also exercising! *Ha!*

Working nine to five (or ten to six in magazine world) proves to be a breeze compared to studying for thirty hours a week and working for thirty more. I can now spend an hour a day on the treadmill until my legs feel like jelly and I can't think anymore. What I've got myself, along with the complimentary doll-size towel from Fitness First, is a brand-new obsession. I swap deprivation for excess.

"Excessive exercise" has been classified as either an addiction or a compulsive behavior. With an addiction, sufferers become "hooked" on the apparent "euphoria" of exercise and so do more and more of it to seek ever greater highs. With a compulsion, sufferers don't necessarily love it, but they feel it's a duty—and they perform it to such an extent that it can cause dysfunction in their personal lives. I am firmly in the latter camp. I don't love it: I just do it.

My therapist is no fool. She can see that I am now not only thin: I'm *sinewy*. Flat-chested, taut and not at all the way nature intended the DD-cupped teenager to turn out. She asks about my exercising habits. I tell her a few semi-truths then add unconvincingly: "I just really enjoy running . . ." She asks if I've ever canceled plans with friends to exercise. "Occasionally," I lie. She asks whether I feel guilty if I don't exercise every day. I don't tell her that I wouldn't know: I never allow this to happen. I realize I may have a problem.

"Exercise addiction is often linked to anorexia," Quinn tells me, "and it's been around for a while." The extensive Obligatory Exercise Questionnaire was created by psychologists from the University of Florida as far back as 1991 and was updated with a six-question inventory created by researchers from Nottingham Trent University in

2004 (available in the Notes at the end of the book).[4] "It can become a real problem and we need to make sure that when we encourage people back into healthy eating habits, they don't feel the compulsion to 'purge' in another way by overexercising," says Quinn. This is sage advice. But I doubt I'd have taken it back then. I keep exercising, daily, until I fall down an escalator.

5

Avoid Excess

You know how when you're a child, you think that perhaps you can fly? I remember trying once, leaping down the stairs four steps from the bottom and landing with a thump on a (thankfully) deep-pile carpet. Well, falling down the escalator into London's dirtiest subway station during rush hour in my hurry to get to the gym after work is a little like flying. For about half a second. After this, it's incredibly painful, humiliating and—temporarily—debilitating. I end up in the hospital having stitches in my shiny black and bloodied shin as a nurse reproaches me, insisting I am "very lucky" and it could all have been "a lot worse." I'm offered antibiotics, told to rest for a day and then avoid strenuous exercise for two weeks.

Two weeks!

This feels like a catastrophe. Worse than the accident itself, even.

What will I do? How will I function? Where will I put all the nervous energy I habitually accumulate if I cannot exercise for TWO WHOLE WEEKS?

Somehow, a day passes following this new advice. It doesn't feel good.

At all. Wine helps, or at least I feel as though it does. I survive. Then I make it through another. I go back to work and people are kind and try not to make me feel like an idiot for pratfalling in such a dramatic fashion. I have friends who keep an eye on me in the most delicate and kind ways: friends who guess that I have a problem and so keep the conversation neutral, sidestepping discussions of food, or weight, or the gym. Friends who keep inviting me out, even though I have withdrawn from group events or family activities for "fear" that there might be food involved, or because I've opted for a night at the gym instead. My friend Steph tells me, "We like you whether or not you're successful in your job or thin!" and it feels like a huge, neon-sign-worthy revelation. Something odd and miraculous. My friend Tony says he'll always be in my corner—and it means the world.

"'I'm here for you and I'm not going to leave,' is one of the best things someone with an eating disorder can hear," agrees Quinn. "It tells some-one that they're worth more than their 'diagnosis' and that just because you *have* a problem, doesn't mean you *become* that problem."

Because I can't exercise, I somehow have the headspace to tune in to the people around me. Instead of running off to the gym at home time, I talk to my colleagues and even befriend a few of them. I so-cialize and have "a life." I work hard to eat semi-normally and get up to a weight that all the medical professionals deem "healthy." But my periods don't come back and I don't ovulate. Loss of fertility is one of the long-term effects of anorexia. This is a worry since I am already starting to feel the scratching of a yearning: a niggling sensa-tion that my body *wants* something. It's a new type of hunger but I know, very definitely, that I want babies. The thought that I might have jeopardized the chance of motherhood in pursuit of "winning" at "thin" makes me shaky. I'm referred to a specialist to see if anyone can tell why I'm still not having periods. No one can. The best ex-planation I get is, "maybe your body has shut down." A nurse, puncturing my arm with a needle to take yet more blood in the pursuit of answers, says, almost to herself: "Maybe it's your heart that's broken." So I drown my sorrows in a bowl of sangria with my new "drinking friends."

I am advised that if I want children, at all, ever, I should "eat up" and start trying for a family sooner rather than later. I am twenty-six and very single. And when I *next* go to my doctor with tonsillitis, I leave with a prescription for penicillin as well as something called fluoxetine. He glosses over the word "antidepressant," but when I get to the pharmacy, there they are. Small. Shiny. Neat. I'm surprisingly attracted to the innocuous-looking capsules.

And so here I am, now, another statistic of a young person with "mental health issues."

I will spend years on antidepressants, the holy trinity of sertraline, fluoxetine and citalopram (more in Chapter 14).

I eat without appetite, stuffing it down me until I feel I've "done the job." I cancel the gym membership and walk more. I develop curves again but am not sure I like the attention they bring. They prompt comments and reactions in people who I wish would just leave me alone, or stop noticing me, like they had when I was very thin.

So I drink. Being anorexic or addicted to exercise won't make you much in demand at parties. But alcohol is a totally acceptable vice. I am nothing if not persistent in my pursuit of a crutch, dogged in my determination to self-destruct. I'm still a lightweight in all respects, but I give excessive drinking my best shot next. To be fair, where I am working at this time, a couple of colleagues are on ketamine most days so this seems fairly vanilla (the Special K Diet takes on a whole new meaning). And it seems to work as a coping mechanism. "You're more fun when you're drunk," is something I hear a lot. It hurts, but I don't like to disappoint. So I get *on* it (early 2000s journalism marking the swan song of professionally sanctioned "day drinking"). Lunchtime drinks with the editor are not unusual. PRs offer generous "hospitality" and socializing with a glass in hand is almost part of the job. "I'm not drinking: I'm networking" is a regular refrain.

I meet a reflexologist at a press event who tells me I'm dehydrated and should lay off alcohol for a while. "Wow, you can tell all that just by looking at my hands?" I ask, amazed.

"No, you just smell of wine and it's 3 p.m."

Oh. Still I drink, with a wretched zeal, because when I drink I am "more fun." I take a chaser after work with the ketamine crew. Because what's the harm? And if five o'clock drinks on a Friday end up lasting until closing time, so what? Who cares if I race to get the last train home and my friend Susie has to haul me back from the edge of the platform before a train hurtles through? A train that would almost certainly have decapitated me had not my friend's white-wine reflexes been sharper than my own. A train whose driver would doubtless have been traumatized by the incident and probably ended up with PTSD. We both could have become another of the statistics on the warning posters dotted around the station. No matter, I think: this is living! This is fun!

Alcohol is known to affect the nerve-chemical systems that are crucial for regulating mood[1] and studies show that depression can result from heavy drinking.[2] Researchers have also proven that reducing or stopping drinking can improve mood.[3] But the drinking culture I come of age in is all-pervasive.

I don't cope well with alcohol, but friends have it worse. One finds that the after-work drinks begin to stretch into two-day binges that cost him his job. Another admits to vodka in his coffee cup at noon. Both are still in Alcoholics Anonymous today (AA has another handy questionnaire to determine alcohol addiction; see Notes).[4] David Nutt, professor of neuropsychopharmacology at Imperial College London stated in 2009 that alcohol was more dangerous than ecstasy and LSD. Other friends have their interest piqued by cocaine, and plenty of people around me spend the majority of their spare cash this way.

The *Guardian* journalist John Crace has spoken openly about his addiction to drugs that kept him hooked for a decade. After experimenting with cocaine and speed, he graduated to heroin when he was twenty, but says: "I actively sought it out—I needed no second invitation. It made me into something I wasn't and I hated being 'me,'" he says. "I would do anything not to be me. 'Not to be me' was completely desirable."

Crace grew up a vicar's son, a role, he explains, that came with expectations. He felt out of place, claustrophobic and under scrutiny: "We were quite bad at communicating but there was a pressure on all of us to be happy growing up. This was immense." His parents both served in the Second World War and were "badly damaged" by their ordeals. His father was sunk twice in the navy and branded a war hero. "But he felt like shit," says Crace. His mother was a Wren and was shot at in Portsmouth. "I think now that they were both suffering from PTSD," he says, "which of course wasn't understood at the time. When they met there was this desire to close things down. To generate a happy family." And so, when Crace was seven, his father left the navy to become a vicar. "There was an understated subtext growing up that it wasn't okay *not* to be happy," he tells me. But he wasn't happy.

Psychologists have found that when we attempt to deny or block a spectrum of our emotions, we can dissociate from ourselves.[5] Dissociation is one of the earliest defense mechanisms to develop (from birth until around age three)[6] and has been defined as a "lack of normal integration of thoughts, feelings, and experiences into the stream of consciousness and memory."[7] If we're taught that being sad is "bad," it makes sense that we dissociate from that feeling. And there is a strong link between dissociation and addiction.[8] If we pursue happiness above all else and become phobic about negative emotions, we're more prone to "anesthetize" ourselves with addictive substances or behaviors. To distract ourselves by "getting high," or "out of our heads," or "numbing" our feelings.

Philosophy professor Peg O'Connor is herself a former addict and author of *Life on the Rocks: Finding Meaning in Addiction and Recovery*.[9] She uses the allegory of Plato's cave as a way to understand addiction and recovery, saying: "There's a temptation to hide or numb out because real life and living is scary and painful. We live in a culture that is afraid of suffering and no parent wants their child to suffer. A lot of parents don't know what to *do* with a child's unhappiness." Many of us have been raised to believe that discomfort is a problem, so when it inevitably occurs we want to "fix it" or make it go away—rather than riding it out. "We've become trained to turn in particular to pharmaceuticals, to give us answers to what I would say are just questions of life," says O'Connor, who adds: "And it's pervasive; to the extent that I'm always amazed when people don't become addicted to something."

On a pinboard above the desk where I now work is a page I tore out of *Stylist* magazine in 2017. It's an interview with comedian, podcaster and former addict Russell Brand where he says: "I was taking drugs and drinking because I couldn't cope with feelings about who I was, so I didn't feel good enough. It was a survival strategy, and I think that's true of anyone doing anything to excess—they can't cope with who they are and they can't cope with the world."

Controversially he also calls addiction "a blessing": "Because if you don't have it bad, you can carry on for ever. I know so many people who wait for the moment before they die to go, 'This isn't who I really was! Ahh, bye!' But me, because I've been smashed on my arse so many times, I've had to do that already. It's a blessing as you have to get real."

You also have to feel.

There's a shame attached to feeling sad in a world that tells us we shouldn't (more of which in Chapter 7). "We internalize this idea that we're somehow 'wrong,'" says O'Connor, "and large swatches of the population have been steeped in shame." Of course, there are many reasons why we develop addictions. "But a lot of people drink and start to use drugs because they feel ashamed or they are shamed by others about different parts of themselves or how they are in the world. And then they become ashamed of their addiction. And it just keeps building and building and building." We get stuck in a vicious circle of shame and poor self-worth. This was Crace's experience.

"I just always felt 'lesser,'" says Crace, "and then I tried heroin and it was the first time I felt 'enough.'"

Interestingly, Crace met his wife during this time and got married, but managed to miss much of his own wedding reception by holing up in a restroom with a dealer. "Our relationship breaks down into two parts," he says, "when I was using and then when I was cleaned up." He finally got clean in March 1987 after a low—celebrating his thirtieth birthday in a grimy SRO with a dealer. With his wife's support, he resolved to stop. "I had to do a lot of work, it wasn't easy making the adjustment," he says. "I was really struggling to have an identity off the drugs and was probably a nightmare. I would say things like, 'My recovery has to come first.' And in some ways this is right. But when there are other people in your life, it's quite selfish behavior."

He joined Narcotics Anonymous, which offered him another source of support and a surprising new focus for his days.

"I had been about a month clean but I was almost incoherent—my brain was somewhere else and withdrawal had taken about three weeks, so I had just started to be sober. Then the opportunity came up for someone to do the tea and coffee—and I stuck my hand up."

To be on coffee duty?

"Yes! I think they were really hoping someone else would volunteer. Someone who'd been clean a little longer. I think that they thought all the tea and coffee money would go missing. They certainly thought at the very least they'd be recruiting a new tea and coffee person in two weeks' time. It was a big element of trust on their part. But I ended up doing it for a year. It gave me a sense of duty and belonging."

The "one day at a time" mantra worked for Crace, partly because thinking longer term seemed impossible. "I would joke with friends that I'd have a big relapse when I turned sixty, because it seemed unimaginable to be getting through the next thirty years without heroin." He's sixty-three at the time of writing and clean, and yet, he says: "The attraction goes on."

Crace knows people who have died from suicide in recovery. He has friends with cancer and heart disease, "and a far higher percentage of people in recovery have mortality-related risks, linked to addiction." I wonder whether he, like many, has swapped in any "healthier" addictions. "Workaholism," he says without a beat. "For the last twenty years I have also exercised addictively. I used to run long distance, then my knees gave way." Now, he spends "hours" on the cross trainer at the gym. "I obsessively collect things, too, like modern studio ceramics. And books. I can get quite into it, then I feel myself withholding and becoming monosyllabic and I have to train myself to talk again."

"I still feel angry about a lot of my early childhood," he says, "the anger hasn't gone completely," although he was able to reconcile with his father before he died. "There was always a sense of 'I don't get him and he doesn't get me'—but I adored him by the end." That's not to say that Crace feels totally at peace ("I'm not some 'higher level being' these days") and he still suffers from depression, seeing a therapist regularly. "I don't expect any of this to go away or get better—therapy for me is more like a form of dialysis. It keeps me going. It will always be there."

It will *always* be there. We are all in recovery. Another friend develops a gambling problem in his twenties, largely because he is sad and doesn't know what to do about it. Research from George Mason and Northeastern Universities[10] found that people who fully experience and work through their emotions are less likely to turn to unhealthy coping mechanisms or experience anxiety and depression. If we allow ourselves to *feel* more, we cope better. But my friend is sad in a world that has told him sadness is not okay. So he stuffs it down. Buries it. Tries to distract himself. Betting starts off as a temporary release—a diversion—but soon escalates into an addiction. He ends up signing away the deeds to his home. Once he finally gets help, he struggles to avoid reminders of anything gambling-related in normal life and nearly has a relapse one day when a mutual friend puts on a DVD of *Ocean's Eleven*.

"Gambling is a lot more common than you might think," says Ian, a former addict turned spokesperson for Gamblers Anonymous. "It's not like alcohol or drugs or food where you can see when someone's an addict: with gambling it can go unnoticed until the money runs out," he says. Ian estimates that more of us know someone—or may even be in a relationship with someone—who has a gambling problem than we realize (feel free to try the "Are You Living with a Compulsive Gambler?" Gamblers Anonymous quiz in the Notes at the end of the book).[11]

Ian first "had a flutter" at age fifteen before developing an addiction that would cost him two marriages, his livelihood and, ultimately, his freedom, resulting in jail sentences. "It's a progressive illness," he says, explaining that illness, marriage breakups and financial problems are common "triggers": "because addiction often happens when we can't show our feelings," says Ian, "and men are bad at that."

I tell him women aren't universally brilliant at it either.

"True," he says. "Perhaps we could all do with getting better at showing it when something's hurting us."

Hear hear.

"A lot of us just tell ourselves, 'Tomorrow will be different,'" says Ian. "But tomorrow can't be different unless we change today." He's lost five friends to addiction ("I've carried coffins. I found one friend hanging from a tree"). Today, he's passionate about pushing for change. "We have to help people get in touch with their feelings, even the bad and sad and ugly ones," he says. There's no data on what percentage of addictive behaviors are caused by mood or anxiety disorders and what percentage are circumstantial—downward spirals that might have been avoided in another environment or by working through emotions effectively. The George Mason and Northeastern Universities study suggests that being able to experience and work through our emotions may reduce our likelihood of turning to unhealthy coping mechanisms. What science *can* confirm is that sadness and addiction are inextricably interlinked.

Addictive behaviors are consistently associated with unhappiness, mood and anxiety disorders, lower well-being, social isolation and stigmatization, according to the latest World Happiness Report. In a chapter dedicated to "Addiction and Unhappiness," the renowned economist Jeffrey Sachs notes that addictions may also give rise to clinical depression: "through mood dysregulation or secondarily to the acute stresses resulting from the addiction." At the same time, depression and other mood disorders can also lead to addictive behaviors, "as individuals try to 'self-medicate' their dysphoria through resort to substance abuse or addictive behaviors."[12]

Social epidemiologists professors Richard Wilkinson and Kate Pickett believe that growing inequality worldwide is leading to a rise in addiction. In their book *The Inner Level*,[13] they write that: "trying to maintain self-esteem and status in a more unequal society can be highly stressful" and "this experience of stress can lead to an increased desire for anything which makes them feel better—whether alcohol, drugs, eating for comfort, 'retail therapy' or another crutch. It's a dysfunctional way of coping, of giving yourself a break from the relentlessness of the anxiety so many feel."

We've all been hurt and struggled to know how to handle it. While heroin addiction may be at the extreme end of the spectrum, there are ample "socially acceptable" addictions that can go unchallenged for a lifetime, like gambling or drinking. We all have our crutches—our unhelpful coping strategies.

"I have regularly seen that it is not the pain of loss that damages individuals, but the things they do to avoid that pain," says the psychotherapist Julia Samuel. It's not uncommon, I learn, to fall prey to addictions or develop reckless behaviors at a time of loss. There's an idea that because we're hurting so much already, we may as well really screw ourselves over.

And yet I wonder why I succumbed so spectacularly to addictions in my twenties, why being dumped by a stupid boyfriend appeared to kick-start a downward spiral.

"A new loss will always bring back a previous loss," Samuel tells me.

"Even if it was just a boyfriend?" I ask.

"'Just a boyfriend?'" She does an eye roll. "'Just' a boyfriend who brought back this idea you had of yourself that you weren't lovable if you weren't perfect?" she asks. "'Just' your first boyfriend who you loved who left you?"

When she puts it like that . . .

There are many triggers for addiction, but mine was born from sadness. Had I been able to cope with sadness better—had I felt that it was *okay* to feel sad—I might not have internalized the depressingly basic route to self-loathing. I speak to friends while writing this chapter and hear stories of excess and deprivation time and time again; coping strategies that distract us for the short term but don't—and *can't*—work. Because we have to feel *all* the feels.

The next step is to get used to this sensation—enough to begin to categorize our experiences into different emotion types. This is a skill called "emotion differentiation" or "emotional granularity" and has been linked to positive mental health outcomes. Although relatively little is known about how emotion differentiation develops in our brains, psychologists from Harvard University and the University of Washington have found that we're pretty poor at it until we're in our mid-twenties.[14] Worse luck.

I don't know exactly what I'm feeling. But I do know that I want a baby at some point, and I'm aware that I may have messed this up by not eating enough for a while and exercising too much. This is big and painful. Too big and too painful to contemplate, in fact (hence the drinking). I haven't got the strength or the emotional tools or the vocabulary to explore what it means if I have just broken my body beyond repair. So I distract myself with another not altogether healthy coping strategy: ricocheting into another relationship.

I am at a New Year's Eve party at a friend's house and there is loud music playing. Possibly by the artist Pink. Several blocks of wood have been pushed together to make a "stage" and a group of us are dancing on it. Well, most of the group is. I keep falling off—because I have been drinking. Most people have drunk much more, so that there is an air of general debauchery and a fair amount of broken glass on the floor. Something prickles on my knees and I keep meaning to look down at them, but then Pink instructs me to "raise my glass," so I do. I know that my shoes should be hurting about now but I stopped feeling pain after drink number one. Eventually, I look down. Everything is red.

Am I . . . bleeding?

I'm ushered off the makeshift dance floor in a fireman's lift and set down next to a roll of toilet paper and someone who starts dabbing at me with disinfectant.

Definitely bleeding.

It appears that I have a fair amount of broken glass lodged in my knees.

Oh well, I think, can't feel a thing!

Once I'm patched up, my carers vanish to take cocaine. I am forbidden. "Not in her state! She's enough of a mess already!" Am I? I don't know. I'll check soon. I feel too heavy right now. So I sit still—or rather, sway slightly, involuntarily. And then across the room, I see a boy I used to know. A boy I was at college with. My friend Tony will later lament: "Are there no *new* boys in London? Why do you keep going out with old ones?" This is a fair point. Perhaps it's because we're drunk a lot of the time and can't see more than a few yards past the end of our noses.

"Hello," he says.

"Hello," I say.

I look at him and he looks at me and I get that old fluttery feeling again.

"You're looking well!"

Uh-oh.

I *feel* something. It's dangerous, but also it means that—just maybe—everything is going to be all right. As though *I* could be all right. That I could be "saved." He dances in a friend's stilettos to show off. He brings me snacks. He speaks softly. He is funny. And charming. And handsome. And very tall.* He drives a convertible and plays Phil Collins in it (What am I thinking?). He likes shiny things, so I decide that I will be his shiny thing. I will "win" at this relationship. I may only have been 90 percent perfect the last time, but this time I will be perfect. What could possibly go wrong?

---

* I'm sorry about this, since several tall friends tell me it's bad form when a short girl goes out with a tall man, as though we're "using up" all the people of expansive growth when we could manage perfectly well with a normal-size one. If it's any consolation: this isn't going to end well.

He pursues me. I let him. He tells me he wants to whisk me away. I let him. I willingly and enthusiastically hand over the reins for my own happiness. To him. Which, as all right-minded independent women of the world know, is ~~a brilliant idea, inspired, definitely going to work out okay~~ a terrible idea.

6

Get Mad

THE SCREEN FADES to black at this point and a Doris Day movie-style montage starts up. A series of rapidly shifting shots tells the story of our relationship, each image replaced by the next in a slow fade. Me: in an assortment of hats and capri pants. New tall boyfriend telling a joke. Me: throwing my head back to laugh (men love it when women laugh at their jokes, right?). New tall boyfriend turning up on my doorstep with flowers. Me: planting a red lipsticked kiss on his cheek. Also me: sliding on sunglasses, patting perfectly set hair, fastening a silk headscarf around my head and setting off on a road trip with new tall boyfriend behind the wheel of a vintage convertible. Albeit one blaring out Phil Collins's greatest hits. The car, now depicted in illustrated form, follows a dotted line around a map of Europe as our 1980s Blue BMW traverses the Continent. Blaring out Phil Collins. Him: handing me an airplane ticket to indicate that our travels have graduated from "minibreak" to "long haul," complete with new, jittery "city music." Me: throwing my head back to laugh (because men love that, don't they?). Him: handing me a glass of champagne on a beach in Hong Kong on our first anniversary (a day-to-view desk calendar flicks by to signpost this momentous date). A series of airplanes take off, to the strains of an upbeat orchestral number featuring a lot of strings, as we vacation in ever more exotic locations. We end on a still of one of these trips—the two of us sucking on straws from a coconut—but as the camera pans back, we see that it's in a frame and it's being mounted on the wall of a new apartment. Tall boyfriend wraps his arms around my waist and stoops to kiss the side of my neck: we are moving in together. I have relocated for him, leaving my job and finding another one nearer his. We're buying patio furniture. And playing Phil Collins. It's enough to make you puke (did I mention the Phil Collins?).

The next scene takes place under weak sunshine at a wedding. It's not our wedding, though there have been hints that it will be soon. We are dressed up in full "guest" regalia. I have whitened my teeth. I have bleached my hair. I have bought new shoes. I have to be "perfect," this time around.

If I'm perfect, he won't leave me, is my exquisite plan.

My rationale is that if I can just get this "marriage" business in the bag, I can get on with the much wanted business of babies, sooner rather than later, as per doctors' orders. And also in accordance with the rapacious broodiness that has taken up residence in my body. Marriage is important to me since my mom and Orange Backpack were not married . . . and he left. Consequently, I deduce that: *if they really like you, they'd better put a ring on it*. I have conveniently overlooked the fact that my mom and dad were married and that didn't turn out roses, either. Marriage has come to symbolize stability in an unstable world—and I want in.

I haven't shared my timeline with tall boyfriend, which, I can now admit, isn't At All Fair. But he's a bright guy. He'll "get it," I kid myself. Like Mrs. Bennet in *Pride and Prejudice*, or anyone from the olden days who's never heard of feminism, my sole ambition at this point is marriage. I am willfully oblivious to any fault lines in our relationship—like the *tiny* fact that after two years he has fallen out of love with me.

We make it through the service, then all move across a field, picking up clods of grass on our shoes, to a tent. After some food, of which I have no recollection but that was probably salmon, there are speeches.

"I just want to say," the groom stammers, sweaty with nerves, "I feel so lucky to have my wonderful bride. She's my best friend and I can't wait to spend the rest of my life with her."

It's all very nice if nothing I haven't heard before. Dancing is next but tall boyfriend doesn't seem eager.

"You okay?" I ask.

"I don't think I can do this," he says.

"Dance?" I ask.

"*Us*," he corrects me.

My face feels hot while my body becomes ice.

It's happening, I think: *it's happening . . . again!*

He's spent the last three months asking me what kind of ring I'd like. He's talked about where our children should go to school. "You even told me to keep the order of service so we could remember what hymns we'd like for our wedding . . ." I pull out the offending article from my clutch to demonstrate.

"I guess I was just trying to convince myself."

Oh. I stop breathing, flooded with shame.

We push through a dance floor of drunken wedding guests to make it out of the tent and back to our bed-and-breakfast. Tall, now *former* boyfriend passes out in his underwear. I lie on the bed in a green silk dress and stare at the ceiling. For hours.

The tall guy breaks my heart.

I experience a pain in my chest and struggle to breathe. "Broken heart syndrome"[1] is a bona fide medical condition that occurs during highly stressful or emotional times, like divorce, the death of a spouse, serious medical diagnosis or times of financial distress, according to researchers from Loyola University in Chicago. Symptoms include chest pain, difficulty breathing and feelings similar to those experienced during a panic attack. The exact scientific explanation for broken heart syndrome still isn't known, but it's thought to have something to do with the release of adrenaline and other stress hormones that have a harmful effect on our hearts. Studies have also shown that a relationship breakup activates the same brain regions that process physical pain. Love really does hurt. It passes (allegedly, although I will still feel a pang of melancholy twelve years later on discovering he married). The pain lessens, with time. But it isn't pleasant. And oddly it's no consolation that we'll *all* have our hearts broken at least once in our lives.

I move out of the apartment I share with the tall guy and into the spare room at my mom's house. It's a long way from my work and I can't afford to live anywhere close enough to my job on the salary they're paying me. So I resign and spend my days circling want ads and looking for somewhere to live while watching daytime TV. I find that where I get to in the viewing schedules before I get dressed is a good indicator of my level of "sad." Watching a rerun of *Frasier*? Why not, for old times' sake. Watching *Murder, She Wrote*? This is fine, I tell myself, because this is my favorite show. I tell myself this because, at the time, it's true. Watching the repeat of *Murder, She Wrote* in the afternoon just to check I haven't "missed" anything? I'm in trouble.

This time, I don't pick myself up, dust myself off and move on, smile on my face. This time, I am a mess—a "grown woman, living back at home with her mother" mess. My mom starts to monitor me via the metric of "days when it's possible to wear mascara without it running." I average one in ten.

I know, logically, that it's no one's fault. You can't help whom you fall in love with—or whom you fall out of love with. But I also can't help feeling angry. Possibly for the first time, ever. I'm mad as hell in fact. I just have no idea how to handle it. How does this even work? The message that "nice girls don't get mad" is one I've imbibed since birth, in common with many raised in similar circumstances. I see men around me getting mad all the time. So how do they get away with it?

"Anger is considered more appropriate in males and so we see in studies that there's a higher tolerance of anger in young boys than there is for girls," explains psychology professor Nathaniel Herr. At best, we womenfolk are allowed to be "spirited." But in general? "Girls and women are socialized to suppress anger—as a feeling as well as a behavior," Herr tells me. This is a problem: "because it's still *there*: you can't deny the feeling of anger. But it's an emotion that often gets underplayed."

Getting mad is normal. It's also been around forever—even Jesus wasn't immune. In the Gospel of John (John 2:13–16) Jesus angrily drove the moneychangers out of the temple because he didn't like them trading in his dad's house (my Catholic education taught me something). What's also been normalized, however, is the idea that women shouldn't "do" mad. Another gem remembered from my school days is the Bible story of Martha and Mary,[2] intended as a cautionary tale about listening to the teachings of Jesus, with the secondary message that if we're female we shouldn't get mad. In need of a refresher? This is the one where sisters Mary and Martha open their home to Jesus as he travels with his disciples. As Martha hurries to prepare refreshments for his arrival, Mary simply sits at the feet of Jesus and listens as he ~~mansplains~~ chats. Martha, noticing this through a serving hatch, in my imagination, then says to Jesus, "Hey! No fair!" Or, in long form and to be biblically accurate:

"Lord, don't you care that my sister has left me to do the work by myself? Tell her to help me!"

To which Jesus replies: "Martha, Martha, you are worried and up-set about many things, but only one thing is needed. Mary has chosen what is better, and it will not be taken away from her."

So while Martha has been working her socks off, Mary is the one getting the props because she was ~~having a rest~~ "listening to Jesus"? Come. On. We *all* know a Mary. ("Oh, sorry I couldn't help with the cleaning up, I was listening to the word of God." I'm not convinced that the son of God comes out of this well, either. Why couldn't Jesus help prepare snacks and make coffee? Why couldn't all three of them talk while tipping a bag of Kettle Chips into a bowl?) Either way, the message is clear: male anger = acceptable. Female anger? Less so.

Today, the psychologist Kimberley Wilson, author of *How to Build a Healthy Brain*,[3] says that it's important to get a grip on our anger, de-scribing it as a "self-esteem emotion" and explaining: "The capacity to be angry really says something about your ability to value yourself." We should, she argues, all be empowered and brave enough to know that our anger is legitimate. And anger, like sadness, can have a purpose.

The neuroscientist Dr. Dean Burnett directs me to a study that shows getting mad can actually lower levels of cortisol. Anxiety and stress trigger the release of cortisol, producing the unpleasant bodily effects that make stress so harmful, but researchers from Germany's Universität Osnabrück[4] found that experiencing anger lowers cortisol, therefore reducing the potential harm caused by stress. Anger can help to motivate us, too. In one study from Utrecht University,[5] participants were shown objects they associated with a reward. Some were exposed to angry faces first while others were not. Those shown angry faces were more likely to strive toward their "rewards." Getting angry can also help us to do better in negotiations. Researchers found that moderate-intensity anger has been proven to result in larger concessions during a negotiation than both "no anger" and "high-intensity anger."[6] In other words: get sad, get *moderately* mad, get stuff done.

The philosophy professor Zac Cogley of Northern Michigan University now distinguishes between "virtuous anger"—think Martin Luther King Jr.—and "vicious anger," which is of no positive value.[7] We have to acknowledge and accept our anger, but we don't have to "lose it." Nor is it obligatory to demonstrate our anger via rage, or "vent." The language used around anger hasn't helped the cause of women's empowerment over the years and has muddied the waters about what the feeling really entails. For years, many of us swallowed the idea that emotions, if not given an outlet, could build up like a pressure cooker and "explode." The "hydraulic theory of emotion," as this is known, has been used historically to justify male violence against women as the "unavoidable" result of men's frustration. "It was a crime of passion," "she had it coming" or "he couldn't help himself" have all been heard in courts of law for centuries. We sometimes hear that prostitution acts as a "safety valve," preventing the outburst of pent-up male desire in rape. But research actually shows that men who buy sex are, on average, more likely to express a preference for "impersonal sex," have greater "hostile masculinity," greater self-reported likelihood of raping and a greater history of sexual aggression.[8] It's also been proven that the idea of withheld male orgasm as "physically dangerous" is a myth.[9] Rather than gouts of unused semen building up somewhere (apologies if you're reading this over breakfast. On with your yogurt parfait . . .), any discomfort felt from being in an aroused state without climaxing is the result of muscular tension around the perineum. And women who are sexually aroused but do not orgasm get this, too (fun fact). A physical hydraulic "release" isn't necessary or inevitable. We can all feel things without having to "vent"—via climax or punching someone in the face. Anger on its own is an energy. And while many men could do with some help in labeling their myriad other emotions so that they don't reach for "mad" when they're actually "sad," most women could do with a crash course in "mad" as a helpful offshoot from "sad."

"Women aren't often great at anger," agrees the Tear Professor Ad Vingerhoets, "especially around conflict." I remember him telling me that, according to studies, women were more likely to weep out of frustration and around conflict since we feel "powerless" and struggle to express anger. This is a theory supported by Herr's research, too. He tells me: "Many women report feeling frustrated when they don't get something they want, rather than 'angry.'"

Regrettably, this rings true. In one morning, I read about several global injustices and then hear a radio report on gender stereotyping reducing girls' willingness to speak up and participate at school. Stories like these make me frustrated and sad, as a rule. But really, shouldn't I get mad? As Audre Lorde, self-described "Black, lesbian, mother, warrior, poet," wrote in "The Uses of Anger": "Every woman has a well-stocked arsenal of anger potentially useful against those oppressions, personal and institutional, which brought that anger into being. Focused with precision it can become a powerful source of energy, serving progress and change."[10]

From Rosa Parks to Gloria Steinem and Andrea Dworkin, as well as the legions of other women in the abolition, suffrage, labor, civil rights and feminist movements, when women get mad, we make stuff happen. Herr argues that we need to encourage future generations of women to get in touch with their anger early on: "We need to teach assertiveness—and model how they're more likely to get what they want if they ask for it."

Of course, many of us have lived with unexpressed anger for so long that turning the juggernaut around takes considerable effort. There's also a cultural bias against women's anger: it's not what "nice girls" do. And if you're a Black woman, the prejudice can be greater still.

I meet the journalist Yomi Adegoke who's written about the clichés of "angry Black women" and tells me that while she, personally, is "very comfortable performing anger," this is "massively limited" for many Black women culturally. "In and of itself, Black women have a right just as anyone else to get angry," she says, "but we still live in the confines of a racist and sexist society, and Black women cannot express their anger as freely and as flippantly as white men." At least, without incurring the label of "angry Black woman." This is a problem, since emotion is a natural part of life. In her book, co-authored with Elizabeth Uviebinené, *Slay in Your Lane: The Black Girl Bible*,[11] she cites research from the US Center for Women Policy Studies revealing that 21 percent of women of color don't feel free to "be themselves at work." The expectation that anyone should "compartmentalize" and suppress anger is not only unrealistic but also damaging. In the book, co-founder of the Black British Business Awards Melanie Eusebe describes anger as "a passionate, driving force . . . a beautiful, healthy emotion that says to us, 'Our boundaries have been crossed.'" She encourages women to acknowledge and recognize their anger and ends with the rallying call: "Do not take away that anger, because there are some things that women should be angry about."

And there's plenty that Black women (and men) should be angry about. In 2020 alone, George Floyd, a forty-six-year-old Black man, was killed by a white police officer who pressed his knee into Floyd's neck for eight minutes and forty-six seconds. Floyd had told police more than twenty times that he could not breathe. Ahmaud Arbery, a twenty-five-year-old Black man was shot to death while jogging after two white men followed him in a truck. And Breonna Taylor, a twenty-six-year-old Black woman was shot by police officers while in bed in her own home.

"I'm just as scared for my kids as my grandmother was for hers," says Jade Sullivan, Black Lives Matter activist, writer and entrepreneur. "Things aren't getting any better. Everything needs to change. We don't even teach Black history properly in schools—this country was built on the backs of Black people and we're written out of the history books." I share that I learned nothing about Black history at school in the UK in the 1980s and 1990s. Sullivan isn't surprised. "But Black history *is* world history," she says. "Like, did you know the person who developed the three-light traffic light was Black?" I confess, even now, that I did not (it was Garrett Morgan). Other Black inventors I've been woefully ignorant of until recently include Frederick McKinley Jones (who brought us refrigerator trucks), Lewis Latimer (carbon light bulb filaments) and Charles Drew (bloodmobiles). I'm still educating myself here: there will be many, many more notable Black figures from history who never made it into the curriculum that most of us were taught in school. "If you want to educate yourself and listen to one thing," says Sullivan, "watch Jane Elliott." This is the anti-racism educator who's been trying to wake up the rest of the world from our ignorance since 1968, when she conducted her "Blue Eyes/Brown Eyes" experiments to teach third-graders about racial prejudice. Something we appear not to have learned more than fifty years later with bias, racial profiling and microaggressions still a daily occurrence.

Sullivan tells me that she and her family have been racially profiled twice in the past week. After being shadowed in a popular women's clothing store by a security guard seemingly convinced Sullivan was planning to steal something, her husband was pulled over and questioned by police. "He was followed around, the police said he 'looked lost' but he was just coming from the post office, home to his family.

The day after we speak, British *Vogue* editor Edward Enninful is racially profiled by a security guard while entering his place of work and told to "use the loading bay." In a post to his one million (and counting) Instagram followers he wrote: "It just goes to show that sometimes it doesn't matter what you've achieved in the course of your life: the first thing that some people will judge you on is the color of your skin."

"The constant pain and sadness, the trauma, the PTSD and the worldwide wounds of racism cannot be measured," says Sullivan, "so yes there's anger. But it's important to have uncomfortable conversations—and to get angry when we should be angry."

We should be angry.

There's a lot to be angry about.

We should all be allowed our anger—and it isn't something we should feel guilty about or mistake for frustration or sadness: we should feel it. And, perhaps, process it. It may not be "nice," it may not be "pretty," and it certainly won't be "comfortable" but it's important. As the psychotherapist Julia Samuel says: "Repressing anger can lead to depression, so it's a better idea to do something with it." Samuel is a fan of kickboxing and advises exercising when we're feeling angry: "You're in fight or flight mode so something like running or cycling, which raises the heart rate and helps ease the feeling of fear, helps reduce stress levels and releases dopamine." She also recommends laughter. This can be a tough one to pull off when you're feeling frustrated and angry. "But it is restorative," Samuel insists. The psychologist and coach Dr. Audrey Tang suggests running through a number of emotions and trying to recognize where we feel them in our body. "This is to get people to be comfortable living with their feelings—positive and negative—so that they can learn to better accept them. So you might think of a time you felt anger—make that picture vibrant—and then ask yourself: 'Where do I feel that emotion?'" Once we know where we feel it, we can start to acknowledge it, be aware and accept. And, hopefully, move on.

I don't know any of this back in 2008. But after a few weeks on the couch, I go a little—forgive the complex jargon—*bananas*. Deciding that action is key, I apply for every job going, somehow securing interviews for a few of them. Unfailingly, I cry *during* the interviews and am then surprised when I don't get the jobs.

Unemployment is no fun. Unsurprisingly, there is a correlation between unemployment, low mood and mental health struggles, with the risk of depression higher among the unemployed. The causal direction of the relationship is not clear—it's possible that unemployment causes poor health conditions, or it could be that having such conditions makes it harder to stay in work.

A Portuguese study found that unemployment depression often impacts men more severely, possibly because men's status has historically been tied up with "providing."[12] But women are not immune. And younger people, who may not have the pressures of a mortgage or caring responsibilities, still experience a significant drop in feelings of self-worth as a result of unemployment. Researchers found the link between unemployment and depression was so strong among eighteen- to twenty-five-year-olds, that they describe it as "a public health concern" in the journal *Preventing Chronic Disease*.[13]

A Swedish study from 2019 confirmed that unemployment makes us miserable and found that unemployment also contributed to a 10 percent drop in health-related quality of life.[14] And that's in Sweden—where everyone's beautiful and the state has things sorted for most unemployed people!

I feel hopeless. And scared. And very, very sad. But instead of allowing these unhappy feelings to sit and feel validated ("I've been dumped! I'm unemployed! I've lost my home! Of *course* I feel bad!") I go into busyness mode again. Fueled by coffee and fear, I go off like a cheap firework. I keep going to stop myself from thinking too much. Again. And I start speed dating.

When I say I start speed dating, I don't mean at an officially organized event: I mean dating as many people as I can, *at speed*. My record during this haywire period is (look away now, family members) a hundred dates in ninety days. *How?* I hear you ask with a hint of horror/wonder (delete as applicable). Well, it turns out that when you're unemployed, free from caring responsibilities and can finally peel yourself away from Jessica Fletcher in *Murder, She Wrote*, there's a lot of "day" to go round. Typically, I get the cheap train to the city a few days a week for a job interview followed by morning coffee (date #1). Then there's lunch (date #2), followed by a drink "after work" (date #3). Never dinner: too formal, too intimate, too much expectation and expenditure. I average nine dates a week, Monday through Friday, reserving weekends for family and friends. With a journalist's training it's amazing how much you can find out about someone in an hour. And while I realize that this can't have been terribly satisfying for the unsuspecting "interviewees," it is a strategy that superficially boosts my tattered self-esteem, fills my days and—controversially—plays the numbers game that dating inevitably is. I don't kiss many of the frogs but I do quiz them on everything from their views on capital punishment to penal reform, Phil Collins ("good or bad?"), and their thoughts on the latest nonfiction bestsellers. Like I say, I am *pretty fun* to date. In what spare time remains, I go out a lot. *Out* out.

I cut loose in a way I never have before in this piratical phase, culminating in a twenty-four-hour bender in Miami on a press trip where, unusually, I'm asked to take the photographs as well as report. Photos aren't my forte, but there's a recession and I'm in no position to turn down work, so I familiarize myself with a friend's fancy camera and set off, feeling like a poor woman's Annie Leibovitz. Journalists don't get paid much but they do occasionally get taken to places they could never afford to set foot in otherwise. This is one of those trips. We stay in the hotel that James Bond hung out at in *Goldfinger* and are plied with cocktails from morning until night. I do not fare well on this kind of regimen on top of jet lag and heartbreak. I click-clack my way around in heels and shorts (it's the 2000s, this is somehow considered acceptable). I struggle not to fall over and snap gamely at anything I think that my editor might like to see. We're shown where James

Bond sunbathed. *Snap*. Where Goldfinger played cards. *Snap*. Where that poor woman was covered in incredibly flattering yet ultimately lethal metallic paint. *Snap*. I get a few strange looks, but think it's probably because they're not used to journalists taking the photos as well.

Perhaps they're intimidated by me as a triple threat, I think: a journalist who can also take pictures and walk (almost) in heels!

They are not looking at me because I am a triple threat.

As I will later discover while trying not to puke on the airplane home, horrifically hungover, they have been looking at me because I've been "taking" hundreds of photographs . . . with the lens cap on. The trip is over. I have no pictures. I am an idiot. Feeling hot and clumsy, I pick up my inflatable travel pillow, take myself off to the tiny restroom at the back of the airplane and scream my rage. Into it. For ten minutes. It's not exactly Michael Douglas in *Falling Down* (yet) but it's as close to an expression of fury as I have ever felt to date. After which, I feel better.

I begin to give fewer fucks. And the fewer fucks I give, the better things start working out. I get more commissions. I get invitations to do fun things with friends I haven't hung out with in a while, because apparently they weren't too hot on the tall guy and all the Phil Collins (who knew?). I take to wearing clothes that are comfortable and soft in a world that is not. And I ramp down the frankly exhausting dating schedule, deciding that maybe I'll give up altogether and become a secular nun in sweatpants. I have one more date in the diary: one set up via a dating site my friend Tony very kindly signed me up to (thanks, Tony).[15] Date number 100 proves himself to be T: a blond Yorkshire man in black square glasses with a wonky index finger, the result of accidentally sawing it off in the Boy Scouts (God bless 1980s health and safety). I like T. I can talk to him. I don't have to be anyone other than "me." And so in the spirit of giving no fucks, I decide honesty is the best policy and tell T that I really want babies, that I don't know if I'll be able to have them and that I've been messed around with before so only serious contenders may apply for the position of "significant other" from here on in. Once he's managed to recover from this bombshell and compose his face into an expression less reminiscent of shock, he nods: "All right then." He wants to see me again. I'd like that, too. So we do.

I kick the tires of this new relationship: it seems okay.

A year later, T and I are grouting a kitchen together (as my mother says: "This one's a keeper!").

Another year later, we're trying for a baby.

This may take some time.

Part II
How to Talk About Being Sad

On shaking off shame and learning not to apologize for sadness; how getting what we want won't "fix" us; and why Russian fairy tales might help us all feel better. From the "Stirrups Years" to living Danishly; love versus grit in relationships; the pain of infertility; the childfree-not-by-choice; fear of failure and what we can learn from it. This is the section in which we explore what the therapist's couch is really like, drugs 101 and why we all need to tune out "Shit FM." Including the tipping point, burn-out, sleep deprivation, diagnoses, the "second shift," boarding school syndrome, trauma, midlife meltdowns, a global wake-up call and how "I am because you are."

Introducing homeschool dad; dream doctor Casper; my mate Jill and the Whirlpool Man. Featuring Nompumelelo Mungi Ngomane, author of *Everyday Ubuntu*; medic turned author Adam Kay; polar explorer Ben Saunders; historian of the emotions Thomas Dixon; journalists Yomi Adegoke, Bibi Lynch and Matt Rudd; Professor John Plunkett; the writer Henry Hitchings; podcaster Marina Fogle; and the evolutionary psychologist Robin Dunbar.

7

Shake Off Shame

L YING ON AN examination table, naked from the waist down
with my legs in stirrups, I have seldom felt more vulnerable. I've
spent several mornings a week in this position for the past two years,
but somehow it doesn't get any easier. This shouldn't come as a huge
surprise—I've been socially conditioned *not* to flash my genitals at
strangers. But still, I am embarrassed and awkward—both by the
ignominy of the situation as well as all that's led me here.

"Sorry," I tell a doctor as she struggles to ram a condom-clad
wand between my stirrupped legs before telling me she can't find
any follicles.

"*After all that!*" she tuts. As though there's been a lot of fuss over
nothing.

"No babies this month," she adds unnecessarily. "Perhaps you're
working too hard."

"Sorry," I tell my boss when I'm late for work, again, because my
hospital appointment overran. I've finally got a job, so that's something.
After riding out the worst of the recession with maternity coverage

contracts and freelancing, I am appointed editor of MarieClaire.co.uk. It's a big, shiny job that I don't want to mess up. But I'm so chock-full of hormones that I have a sneaking suspicion I'm sabotaging this, too. I'm also embarrassed. Ashamed that my body isn't working the way I want it to and pained by the language of "failure" used about it. About me. About the fact that I have "failed to respond" to treatment. Again.

I don't tell my co-workers what's happening but I try to talk to friends about it when they ask why I'm not drinking ("Where's 'fun Helen'?"). I explain that the Clomid—a drug I'm taking to stimulate ovulation—makes me nauseous. Next, that the follicle-stimulating hormones cause vomiting, diarrhea, bloating, pelvic pain and acne. My fridge at home becomes filled with vials of liquids and preloaded hypodermic needles. I carry a mini cooler of medicine around with me, in case I need to administer on the go. I inject myself in airport restrooms, train stations, at the office and, once, backstage at New York Fashion Week (the glamour).

"Sorry," I tell T, now my husband, when he has to inject HCG "trigger shots" into my rear to stimulate the release of an egg because my own hands are too shaky and slippery with sweat.[1] He rubs at the stubble on his jaw, as he does when he's anxious, and rolls up his sleeves as though he's a dairy farmer about to deliver a cow. "Is this how you imagined married bliss?" I ask.

"It is not," he admits, sticking the needle in.

People around us get pregnant. Some, swiftly. "Honeymoon baby" is a term I hear with increasing regularity. *Congratulations!* I smile, then cry. I am grateful to the ones who send a text or email to let me know their news before I see them face-to-face, allowing me to compose myself before we meet in person. This way, I can devote my energies to being happy for them. Because I am happy for them. I just wish it were happening for me, too.

While friends talk about the best Baby Gap booties or how to make a rainbow cake, I become obsessed with signs of an "egg white" consistency in my underwear—the goal of all "trying" couples everywhere since it's a good indication of ovulation. Whenever an email goes around the office that we're about to present the "card 'n' cashmere baby blanket" gift basket to whoever's off on maternity leave this week, I grind back my chair, pinch the fleshy part of my palm so as not to cry and stuff snacks in my mouth to avoid having to say anything. I'm getting surprisingly good at "misery eating." I try to be front of the line for goodbye hugs with the mother-to-be so I can get the whole thing over with. Then I go to the furthest cubicle of the restrooms on the third floor and the tears come. I am stretched—bloated with hormones that are not my own—so that some days I look as though I'm six months pregnant. Only I'm not.

Between one in six[2] and one in eight couples[3] in the US have difficulty conceiving. I'm nothing special. But I am very sad about the idea that I might never hold my own baby in my arms.

Studies into the psychological impact of infertility show that the process of trying and "failing" repeatedly takes its toll—on all parties. Research from Middlesex University found that 90 percent of respondents struggling to conceive reported feeling depressed, with 42 percent feeling suicidal. More than two-thirds said that infertility had a detrimental impact on their relationships, with 15 percent admitting that their relationships ended or became strained due to fertility problems. Oh, and half of the women and 15 percent of the men surveyed said that infertility was "the most upsetting experience" of their lives.[4]

Of their lives.

The journalist Bibi Lynch has spoken openly about the pain of not having children and that despite having watched both her parents and her uncle die and having been around violence and alcoholism—"not having kids is without doubt the worst thing that has ever happened to me." I first read about Lynch's childless grief in a newspaper in a hospital waiting room ahead of a "debrief" on why my body had "failed to respond" to another round of treatment. I never forgot her words, though I remember little about the rest of that day. Lynch is unflinchingly honest about how she came to join the ranks of the "childfree-not-by-choice" (or CFNBC) and so I'm eager to speak to her while writing this chapter.

"I just always wanted the whole package," she tells me, "to be with someone and have a kid with them. But I didn't meet anyone. 'Social infertility' they call it." As the eldest of seven siblings, she had always assumed that she would have a family and that motherhood was "just something that was going to happen. But it didn't."

"I was conned into thinking I had more time," she says, "duped by all the coverage of 'miracle babies,' women who conceive later in life, or a Hollywood A-lister who had a kid at forty-eight. So in my head, I thought: I could still do this!" Then Lynch's father died. "Grief makes you think of the whole circle of life," she says, "it really focused me." She decided to try having a child on her own, aged forty-two. She bought sperm, paid for follicle tracking—where they check ovulation—had scans and blood tests. "It cost thousands," says Lynch. To go through with the final stage of IVF, follicle implantation, would cost a further $4,000. But then the 2008 recession hit and she lost her apartment. "I didn't have enough money and it was . . . awful."

The grief she describes is devastating. "But what's worse is that in our society, you're not allowed to grieve for the child you didn't have—the future you don't have," she says, "it's not accepted. So your grief is disenfranchised, too." Lynch isn't sure why this is ("I don't know why anyone's grief would be challenging to someone else's") but says that it clearly makes people "uncomfortable." Which isn't much fun for her, either.

"So there's this visceral grief of not having a child when you wanted children," she says, "and there's the grief of not having that love that society tells you is 'the reason to be alive' . . . and then there's also a feeling of having to find your own way. Because otherwise, what's your role? Where do I fit in society when I'm seventy as a woman with no children? I don't think you *should* be judged by that," she clarifies, "but you are." Then there are the painful things that other people routinely say to the childfree.

Oh yes. These.

"Why don't you just adopt?"; "Have you thought about surrogacy?"; or "Just relax more!" are standard suggestions I get during this time. To which I answer, respectively and respectfully, "No, because of reasons"; "No, see: 'reasons'"; and "Gosh, yes, good one. I hadn't thought of that." I *think* people mean well. But stories of miracle babies conceived by moonlight, eating watermelon, facing west, do not make me feel better and do not make me pregnant.

"I'll be told, 'Have my children for a weekend—that would put you off!'" says Lynch. "I've had someone insist that he was more exhausted than I was because he had children," she says, "at a time when my uncle was dying, I was suffering from severe depression, I was homeless and I wasn't sleeping. I just felt like, Why won't you let me have my exhaustion?" Lynch tells me that she was once approached at a funeral and asked: "'Did you get married? No? Did you have children? No? Oh, well. It's not for everyone . . .'" This is quite the sucker punch.

And how does she respond when people say things like that?

"Either I'm exasperated and I'll eye roll, or I'll say something. And then that ruins the day because I've 'said something.' So I have to suffer grief while having to defend my right to grieve. I'm attacked for doing so, and then I'm expected to be the one who apologizes!"

Lynch is now fifty-four and only stopped paying storage fees two years ago for the frozen sperm she bought. "It's devastating," she says, "but I'm working on it." By "it" she means the grief. "And maybe there is other stuff in life," she says. "It's not going to be the same, but maybe it's as worthy, as fulfilling. I'm getting fiercer about it all. I'm not going to apologize for my grief."

Anyone who wants to have children and can't is likely to experience profound sadness. I speak to Richard Clothier, who experienced male factor infertility and now speaks out to break the stigma attached to infertility. He tells me: "My wife and I got married the day after Prince William and Kate Middleton, so we'd joke that the race was on to see who could get pregnant first." But after two years, nothing happened. "Our doctor said that my sperm was a little slow but to keep going and that we'd be pregnant by Christmas." December came and babies didn't. "Eventually, we saw a locum who told us, 'I'm so sorry but with these numbers you're not going to conceive naturally.'" Hearing this was a shock. Clothier started to look into what the diagnosis meant and learned that he wasn't alone. Male factor infertility is on the rise and the sperm counts of men in Western countries have halved in the past forty years, according to 2017 analyses.[5] Studies into the experience of male factor infertility are relatively scant, but what quality research there is has found that it's often viewed as a "failure" of masculinity—a shaming experience that's both isolating and traumatic.[6] "There was a real guilt attached for me," says Clothier, "because the reason for the infertility was mine and I could see how much the lack of a baby was affecting my wife." He and his wife started IVF, but the first cycle of treatment didn't work ("We found out on Mother's Day"). He confided in one friend: "But then *his* wife got pregnant shortly after. When I congratulated him on the pregnancy, he shrugged and said, 'I'm just glad to find out I'm fully functional.' So that hurt." After that, Clothier closed himself off: "I didn't have anywhere for my sadness to go. I felt I had to support my wife, so I largely buried my feelings."

His one sanctuary? The car.

"There was this stretch of road on the way to work, and I would always do 'my sad' there. I'd get really upset, crying, during really tough times when I felt I couldn't show that sadness at home or at work, or with my friends, or anywhere. I remember once driving in the snow and that experience when you turn the wheel but the car keeps going straight. I think acknowledging the sadness is a little like turning into a skid. The equivalent of acknowledging what is happening." I like the analogy of sadness being like a skidding car. Motorists are supposed to turn the steering wheel into the direction of the skid, since braking will lock the wheels, making the car skid further.* To get out of both a skid and sadness, we have to accept what's happening and turn toward what scares us, rather than away from it. It's counterintuitive, but necessary. And it certainly helped Clothier. After the first round of IVF failed, he and his wife saved up and prepared for a second cycle of IVF by making plans: "If it didn't work out, I thought we'd have to move. We live in a small town and we felt it would be too much when others around us started to have children. I thought we'd build a new life for ourselves, actively seek out the company of other childless-not-by-choice couples. So that at least we knew we would be in 'safe' company. Where the people around us wouldn't suddenly get a 'surprise' or 'good news.'" News that might be too painful to bear. "I somehow needed to get something down on paper," he tells me, "a protocol to trigger if or when it didn't work out."

He also realized, finally, that he shouldn't have to feel shame or guilt or apologize for how he was feeling and so, rather than hiding his sadness, he went public with it. "I'd had enough by then. I thought, Why aren't people talking about this? so I wrote to media, my MP, everyone." He expected this frantic period of letter writing to be a release. "What I didn't expect was people to take me up on it." But they did. Clothier spoke on TV and radio—an experience he describes

---

* Kevin the driving instructor didn't teach me this: the winter of 2000 did.

as cathartic. "One friend heard me on the radio and texted: 'I had no idea you were going through this. If you want to go and walk up a mountain together and scream and shout, I'm here for you.' I loved that. Because really, the experience of infertility had blindsided me. It felt like bereavement."

Anne Chien, infertility counselor, agrees that this experience is upsettingly common: "It is known that high levels of emotional distress can be experienced by individuals who are having fertility treatment and that it can impact their relationships with partners, family and friends." She encourages counseling as an integral part of fertility treatment—but this is easier said than done, with professional talking therapies difficult to access for many. Instead, the childfree-not-by-choice tend to get their advice from amateurs—asked for or otherwise.

I plow on, weighed down by a strange sort of shame—lugging it around with me, like a deeply uncomfortable, poorly designed, in no way ergonomic or attractive backpack. The backpack of shame. For this chapter, I spend a few weeks deep-diving into shame, retrospectively. I learn that shame, although unpleasant, serves a function—at least, in evolutionary terms. As with all negative emotions, shame has a point. In what's known as the Social Self Preservation Theory, situations threatening our social value or standing—like flashing our genitals at strangers or failing to have babies when society thinks we should—will *decrease* our self-esteem and *increase* cortisol levels as well as feelings of low social worth.[7] In other words: provoke shame. Shame is *intended* to defend us against the "social devaluation" that can come about when negative information about us reaches others. It does this by deterring actions that would lead to devaluation[8]—like flashing our genitals to strangers. It's supposed to help us play nice and not flash our genitals at strangers. So I feel shame when naked with my legs in stirrups because society has instilled in me the importance of *not* flashing genitals at strangers. Unfortunately, I have to flash my genitals at strangers to try to have the babies that society (and I) also want. Confused? Me too (I blame all the infertility drugs). Unfortunately for many of us, the natural shame impulse can go into overdrive.

Carl Jung, Swiss psychiatrist, psychoanalyst and founder of analytical psychology, called shame a "soul-eating emotion" and whereas guilt revolves around the feeling of having "done something bad" shame is linked to the feeling of "I am bad." Shame is something that makes us see ourselves as defective or worthless and—as we discovered in Chapter 3—shame has strong links with addiction and eating disorders. There's also a high correlation between shame and depression, violence, aggression, bullying and suicide.[9] Shame affects us all, but the way it wheedles its way into our lives and the times when we're most susceptible are often gendered.

When *Elle* magazine published "4 Times a Woman Is Mostly Likely to Experience Shame in Her Life" in July 2015, friends and I responded with a good deal of nodding, the odd "*uh-huh*" and a fair few "*you don't say*'s." The journalist Victoria Dawson Hoff asked a panel of women's mental health professionals to pinpoint the times in a woman's life where she might be more prone to feeling ashamed and, unsurprisingly, the teenage years came out at number one: "a cesspool of hormones and raw emotion." Being made to feel ashamed or too pushy or not "good enough" in the workplace was at number two, followed by postpartum shame, when there's a pressure to be the "perfect" mother. Finally, the fourth most common life stage identified was being thirty-something, single and experiencing the pressure to pair off and reproduce.

With the life experience of another half decade or so since the piece came out, I'd add others. Trying and "failing" to conceive doesn't feel great. Relationship breakdowns, family estrangements and abuse can also induce unwelcome shame. And then there's the trauma of miscarriages—something many friends have suffered in silence in recent years. It's estimated that as many as one in five pregnancies end in a miscarriage, according to Mayo Clinic data,[10] but few who lose a baby feel "allowed" to properly grieve the loss of their baby or the life they'd hoped for. Nearly 20 percent of women who experience a miscarriage become symptomatic for depression and/or anxiety, according to 2015 research published in the *Primary Care Companion for CNS Disorders*.[11] These symptoms typically last between one and three years, impacting quality of life as well as subsequent pregnancies. And yet, miscarriage is seldom spoken of. "Because of this silence, people don't realize how traumatic it is—until it happens to them," US journalist Hadley Freeman wrote in the *Guardian* in 2017. "I certainly didn't."[12] Silence is often toxic and leads to shame. The writer Christen Decker Kadkhodai described her miscarriage, also in the *Guardian*, in 2016, saying: "It does physically hurt, but what hurts more is the shame. I feel ashamed, small and powerless."[13] Shame is everywhere. But it seems that the way we experience it depends on the body we inhabit.

The researcher and writer Professor Brené Brown argues that the messages of shame in our society are often organized around gender: "For women there are whole constellations of often contradictory expectations that, if not met, are sources of shame. But for men, the overarching message is that weakness is shameful. And since vulnerability is often perceived as weakness, it is especially risky for men to practice vulnerability."[14]

James Mahalik and his colleagues at Boston College conducted a study where participants were asked: "What do women need to do to conform to female norms?" The top answers were all about being "nice," "thin," "modest" and using "all available resources" to enhance their/our appearance.[15]

Good grief . . .

When Mahalik asked about what men need to do to conform with male norms, the answers were: "Always show emotional control, work, pursue status and violence."

This is deeply depressing though not surprising. A friend who was in an abusive relationship became consumed with shame about what people would think if she spoke out—if she stopped covering up the bruises. She worried that others would view her in the same way *she* used to view women who chose to be with men who hurt them. She worried that she would be judged in the harshest terms and so marinated in her shame for years, staying silent. Shame held her back—as it does so many suffering from domestic abuse. At the time of writing, the UN has described the worldwide increase in domestic abuse as a "shadow pandemic" alongside COVID-19. It's thought that cases of abuse have increased by 20 percent during the lockdown, as many are trapped at home with their abusers.[16] This is horrific. Violence isn't a private matter and it isn't a "choice." We need to realign our idea of what's "shameful" as a society.

There's something called "normative male alexithymia" that plays a role in characteristically "male" shame. Alexithymia is defined as the "inability to identify and express or describe one's feelings"[17] and around 10 percent of the population suffer from this. But the psychologist and former president of the American Psychological Association, Dr. Ron Levant, came up with the phrase "normative male alexithymia" decades ago to describe the inability of some men to put emotions into words thanks to traditional masculine role socialization. In a nutshell, the formation of "masculine" identity conflicts with the emotions men feel and believe they're allowed to express. Traditional "masculine norms" are endorsed and conformed, too—"emotions" don't get much of a look in. So that when emotions inevitably seep out (because: natural, normal, human) there is shame. Big shame. The socialization of men to fear showing vulnerability or "weakness" is so pervasive that some degree of alexithymia is apparently now "the norm" for most men (hence, the "normative" part).[18]

Levant has investigated the link between normative male alexithymia, relationship satisfaction, communication quality and fear of intimacy. In a 2012 study he confirmed that, lo and behold, normative male alexithymia was indeed linked to lower levels of relationship satisfaction, lower quality of communication and a greater fear of intimacy.[19]

According to Brené Brown's research, the antidote to this shame is vulnerability. Vulnerability, she argues, is far from a weakness: it is strength. Courage, even. We shouldn't feel ashamed of our emotions and we shouldn't apologize for being vulnerable. Ever.

We *should* apologize if we've done something wrong. We just shouldn't have to apologize for *feeling*. And yet many of us often do.

8

Stop Apologizing for Feeling

BACK IN MY flashing-genitals-at-strangers period (what I'm calling the "Stirrups Years") I notice friends and acquaintances apologizing for their feelings on a daily basis. Around this time, one of my team members arrives at work puffy-eyed.

"Are you okay?" I ask.

"Yeah," she says, then: "No, my aunt died."

"Oh, I'm so sorry!"

"No, *I'm* sorry." She presses her eyelids. "It was a year ago . . ."

"Oh!" I know by now that there's no time frame on grief. But she goes on.

"My mom told me last night."

"Your mom only just told you that your aunt died?"

Colleague nods.

"Wow. Were you and your aunt close?"

"Well"—colleague hesitates—"we hadn't spoken in about a year."

"Right. No." This makes sense. "I'm *so* sorry . . ."

"No, it's fine. *I'm* sorry. Sorry for getting so emotional . . ."

What? Someone died! And no one told you?! And YOU'RE sorry? What kind of crazy world are we in where we're apologizing for death?

Another colleague tells me that her boyfriend has just been diagnosed with cancer. I send her home to be with him.

"Thanks," she tells me, then: "Sorry!"

You're sorry your boyfriend has cancer? Or you're sorry you can't work today? Either way: stop apologizing!

Another friend gets knocked off his bike by a car. His kneejerk response? "Sorry."

The word "sorry" has been in use in one form or another since Anglo-Saxon times. *Sarig* in Old English is an adjective that means "sorrowful"; so a state of being—an emotion, even—instead of a mere apology. "Sorry" in the traditional sense could be used to signify our acknowledgment of existential angst, the inevitability of mortal pain and the ultimate futility of existence. But today, it's employed as though expressing remorse for *feeling* sad rather than for the sadness itself.

Because sadness is seen as somehow embarrassing. "Shameful" even. And North Americans and the British are particularly culpable here.

The Booker Prize winner Julian Barnes wrote about the loss of his wife, the literary agent Pat Kavanagh, in *Levels of Life*[1] and describes how he would talk about her in conversation and friends would fail to respond: "Afraid to touch her name, they denied her thrice, and I thought the worse of them for it."

The British author Henry Hitchings encountered a similar response from friends after his mother died and tells me that the most candid responses came from Latin cultures. "A Portuguese acquaintance, for example—someone I hadn't been terribly close to before—said some really heartfelt things."

And how did that make you feel?

"Strange," he admits, "but not uncomfortable. It felt very *human* and quite surprising—because we are bad at sad. We don't like to appear vulnerable. Take sports . . ."

This isn't where I'd been expecting the conversation to go, but okay then.

"When I was growing up, it was something coaches drummed into me: 'Don't show any pain.'"

"This seeps out to other areas of life," says Hitchings. "Don't show weakness—the idea that you are ceding some psychological ground if you show weakness: giving the opposition an advantage." He believes that, as a sporting nation, a good percentage of the British population may have been raised with this rule of the game: "The idea that it's disadvantageous to show any signs of vulnerability gets bred into us." Especially when it's sadness.

Viewing sadness as an embarrassing imposition seems particularly punishing since it occurs at times when we need support more than ever. Sadness is normal. Grief is normal. And we shouldn't apologize for our emotions.

"There is this pressure we put on ourselves when something sad happens to somehow pretend that everything's all right and carry on as normal," says Marina Fogle, prenatal teacher and host of *The Parent Hood* podcast. "But life can't carry on as normal after something really sad has happened. And we should be able to talk about this without feeling the need to apologize." Marina and her husband, TV host and adventurer Ben Fogle, suffered tragedy in 2014 when their son Willem was stillborn. When hospital staff told her that her son had died, she went into shock at first. "It didn't seem real: I held him in my arms and just felt . . . numb. It wasn't until the third day that the tears finally came and I wept as though my heart was broken—which it was."

In the days and weeks that followed, Marina found that not only was she grieving, she had to tell other people about what had happened and cope with their reactions. "I remember one woman I knew asking if I'd had my baby yet. When I told her, 'No, he was stillborn,' she went white and I sort of supported her, physically, and *I* apologized. My baby had died and *I* was feeling guilty for ruining this woman's morning—for making *her* 'uncomfortable.' It was ridiculous!" She experienced the strange and perverse shame that many of us attach to grief and feelings of profound sadness. Marina went into "doing mode" soon after and went back to work at the prenatal class she runs.

"I didn't know anyone who'd lost a baby before. And death wasn't something that anyone I knew ever talked about—or if they did, they'd apologize for it. As though talking about anything too upsetting wasn't allowed. As though talking about it might make it worse. But talking about the loss *can't* make it worse—because the worst thing has already happened. It's not as though I'm walking around *not* thinking about it. It's not as though I could ever 'forget' that Willem had died. So talking about it isn't going to 'remind' me of his death: it's always with me. Talking about the loss can only help." Marina learned how to start talking and stop apologizing with the help of the psychotherapist Julia Samuel and grief counseling: "I learned that it's not my job to stop other people feeling uncomfortable about what's happened—or to say 'sorry' for my sadness."

This feels like something we could all do with getting our heads around. Of course, there are degrees of loss and pain and sadness. But we will all be beset by loss and sadness and we can't avoid feeling sad when we need to feel sad. Pain is still pain. And we can have empathy for others while also cultivating empathy for ourselves. "It doesn't make our sadness less legitimate," as the child and adolescent psychotherapist Jane Elfer puts it, "and we shouldn't feel ashamed of feeling sad."

Within families, a "ranking of grief" often emerges to ascertain who has priority in grieving, and who has to "defer" their grief. But the late psychologist and Berkeley scholar Harvey Peskin asserted that although this "ranking" was commonplace, it shouldn't be, since the "right to mourn" is a fundamental *human* right.[2] There is no affidavit of authenticity on sadness—there's plenty of it to go around and we all get to feel it when it comes calling.

I'd like to say that I come to this realization by myself thanks to my inherent wisdom and insight. But what actually happens is that I become weary and sputter to a halt, one wet Wednesday.

I make it home from work after another day hauling around my "shame" backpack and shut the door behind me with relief. I peel off painful shoes and start foraging in the refrigerator for dinner. I am in this position, head in a cheese box, when T walks in and tells me he's been offered a job in Denmark. A recruitment agent has been in touch, on spec, and offered my husband his dream role working for the Danish toymaker, Lego—not in Copenhagen (the fun place that people are always saying is "wonderful") but in rural Jutland.

"Rural Jut-where?" I hear you ask. To which the answer is: "Exactly."

At this point in time we have never been to Denmark. We cannot point out Denmark on a map (*of* Denmark). So we visit for a weekend, just to check the place out. Everything is green and clean and empty. People look more relaxed than we're used to. They take their time to stop and eat together. Or talk. Or just . . . breathe. And we're impressed. T is sold on the idea and begs me to consider.

I am tired after twelve years in the city at this point. My pincushion of a body is broken. After monthly disappointments, my soul could do with a sabbatical, too. So I say "yes." I resign from my big, shiny job, and we emigrate, leaving the bright lights and bustle of a major city for rural Jutland, in the middle of winter. I don't know anyone, I don't speak the language and T leaves for work at 7:30 a.m. and I'm all on my own. I hammer out emails to editors, begging for work, and then I walk. For hours. Through forest that looks a lot like a scene from *The Killing*. Trees tremble with snow, occasionally sliding great globs of the stuff onto my head. I climb steeply up unfamiliar banks in search of footpaths and find myself sheering across the ice, tumbling down the other side on my ass. After getting lost for anything from a few minutes to a few hours each day, I typically emerge into a world shrouded in grey (welcome to Scandinavia) and see columns of cloud roll in from the sea. I watch the weather until my nose turns numb from cold, then go home. Often, via a bakery. Bakeries are abundant in even the smallest towns in Denmark and the pastries are delicious. So I tell myself that this is an important part of "cultural integration."

I start writing about this brave new world and the Danish way of life for a national newspaper with an irreverence that is new for me. Embracing radical candor, I stop apologizing, shake off shame and find my voice. Giving fewer fucks suits me and I am asked to write more. I write as though my skin is flayed open, unashamedly, for the first time ever. And then someone tells me they want to turn the words into a book.

This is officially the best news I have ever heard.

Halfway through writing what will become *The Year of Living Danishly*, I notice that my hair has turned into more of a lion's mane and I look a lot like Aslan, from *The Chronicles of Narnia*. I've also been feeling sick and my DD breasts are back and painful. I Google these "symptoms" and the internet tells me that I might want to think about peeing on a stick. So I do. Then, disbelieving the result, pee on four more.

I'm pregnant.

This feels like a proper Jesus, Mary, Joseph and all of his carpenter mates' style *miracle*. But a part of me also feels guilty for "defecting" to the "fertile side." I start apologizing for deserting the cause and joining the #AsAMother club. How can this be? I remember how it felt, when friends I'd thought were in my "crew" had a change of fortune. So I'm cautious.

Part of me becomes convinced that it "won't work" or that I won't be able to carry to full term. Babies, I know from painful firsthand experience, are fragile. Babies, I know, can die. Some don't even get to take their first breath, I remind myself daily. But despite the odds, I stay pregnant. For ages.

In week forty-two, I am induced and a sickly throbbing transmutes into an agonizing, barbaric disruption. There is no easy way to get a baby out and the top half of my body divides and floats off. There is agony, for eighteen hours. The midwife has time to knit a woolly hat during my labor. Students come in to take a look. Then, finally: a mewing sound and a small, rubbery thing is placed on my chest momentarily before being whisked away to the special care baby unit.

I'm sorry, I'm so sorry, I'm so, so sorry . . . I scold myself through a fog of pain: How could I have dared to think I could do this? Conceive and grow and deliver a healthy baby?

Finally I'm pushed in a wheelchair to see my baby—a boy. He's alive. He's okay. I want to tell him again that I'm sorry: that I've messed this up. All the books emphasize the importance of a natural, calm, tantric-yoga-fest of a birth followed by instant skin-to-skin contact, and I haven't even held him yet. I know all about the benefits of breastfeeding but, instead, he's hooked up to a feeding tube. And a breathing tube. And instead of "me" he's holding on to a small, yellow . . . *octopus?* I look around. Every other baby in the special care baby unit also appears to be clutching a crocheted octopus. A nurse explains that babies squeeze their tentacles to remind them of the umbilical cord in the womb, to calm them and help with breathing. "The babies are also less likely to pull on their tubes if we keep their hands busy," the nurse tells us. In all my tentative psychedelic dreams of motherhood, I never once envisaged *tentacles* (more fool me).

Finally, I get to hold my son and my heart triples in size. He has flame-red hair (no one knows where that comes from), a scrunched-up red face and some impressive lungs. He is given the all-clear after three days but I'm kept in the hospital for a week. The birth is classed as "traumatic" and my body apparently recoils so much from the experience that it tries to go into menopause and stops producing estrogen. Attached to drips, monitors and a "drainage bag," a commode is eventually wheeled in.

"What's that?" T asks innocently.

"Will you tell him or should I?" asks the nurse with an arched eyebrow. It's lucky I've had some practice in shaking off shame, I think. In the past twenty-four hours I've doubled the tally of people with intimate access to my nether regions and now I'm expected to defecate in front of an audience. But none of this matters: we have a baby.

9

The Fallacy of Arrival

For our first Christmas together, T gave me a personalized tree ornament that read: "Book & Baby." He gave me this, first because I'm boring to buy for (I like books, bookstore gift cards and audio books) and, second, because these two things were all I wanted. Books and babies having been the sole aspirations of my adult life. We all have something like this. It could be getting to a certain rung on the career ladder, or earning enough to stop having to do math at the grocery store. Or meeting the partner of our dreams. Or having sex with several hundred partners of our dreams. Or achieving fame as well as fortune (and all the sex). There will always be something that we're convinced will make us "complete."

Having a baby and being a part of the magical world of "books" were all I'd hoped for. And now I had these things. I *had* my bauble! I'd never be "sad" again!

Except when I would.

Because, sometimes, I could.

Because life sucks that way and we are all mere deluded mortals.

I am well aware of the irony of moaning about not having a baby and then moaning about having one. But thanks to a tricky birth and my body's subsequent trip into "a menopausal state," my stitches don't heal for three months and I keep having to go back to the hospital, to be cauterized. *Internally*. The smell of burning flesh will never leave me.

My baby cries a lot. His face crumples up, puce. His fists are permanently raised in tiny balls and his surprisingly strong limbs thrash about.

"He's a fiery one!" friends half joke.

"Maybe it's the red hair!" is something I hear. A lot.

"Can you get angry babies?" I ask T.

"Apparently so." He holds a hand to his face where he's just been punched.

"Oh."

I understand that we are all a genetic lottery. It could be that this is just who he is. But it's probably me, I think, my fault. If our personalities and predispositions are down to genetics, early experiences and lifestyle, then parents can potentially be blamed for most things, I rationalize.

I conclude, variously, that my baby must be angry because I didn't do enough hypnobirthing practice. Because I didn't do enough yoga. Because I was too stressed during pregnancy. Or before, even. Researchers from King's College London have found that maternal stress before and during pregnancy can affect a baby's brain development[1] and a study from the University of Bristol showed that children of anxious mothers were twice as likely to have hyperactivity in adolescence.[2] *Namaste!*

I'm placed in a Danish mothers' group by my health visitor to "get me out of the house" but find that all the other moms have deliciously fat, docile creatures. Their babies sleep contentedly, waking only occasionally to breastfeed successfully, before passing out again in satiated bliss. My baby is very much of the Gordon Gekko school of thought in his conviction that "lunch is for wimps" and snoozers are losers. Enticing him to feed while he writhes and turns his tiny head from side to side is a lot like playing a magnetic fishing game: trying to catch carp while they rotate and mechanically whirr, snapping mouths open and shut at random.

"How can it be this hard?" T asks. "People have been having babies forever."

I shake my head: I have no idea.

"Maybe people have been miserable forever: they just didn't talk about it," he mutters.

Several friends from home helpfully refer me to a YouTube spoof entitled *Danish Babies Don't Cry*[3] set to the tune of Bruno Mars's "Uptown Funk." I learn that Denmark, as well as being one of the world's happiest countries, also has the most contented babies ("braggy fact"). According to a meta-analysis published in the *Journal of Pediatrics*, Danish, German and Japanese babies cry the least, while British and Italian babies cry the most. Danish babies also displayed the lowest levels of colic in the study and many more Danish mothers breastfeed than elsewhere.[4]

This is thought to be because new mothers in Denmark are less stressed and have more time on their hands, thanks to generous parental leave and living in close proximity to extended family—another good indicator of happy babies and happy mothers. Unfortunately, as a self-employed Brit living in Denmark, far from anyone I'm even remotely related to, a tight-knit familial support network isn't an option and I do not have a happy baby. I have a baby who refuses to sleep or eat and shouts at me all the time. And then my milk dries up.

I've blown it! I think. He's only three months old and I've blown it!

Approximately ten thousand newspaper articles tell me that failure to breastfeed until a child is at least of voting age is subhuman, so I beat myself up over this, too. I have a brain that's been trained to tear things to shreds, including myself. I'm temporarily comforted by research conducted by the University of Liverpool[5] showing that mothers experience negative emotions—guilt, stigma and the need to defend their feeding choices—*regardless* of how they feed their baby. For the first but by no means the last time, I realize that mothers can't win. They also can't sleep.

I read a news headline that shouts: "Parents of newborns miss out on SIX MONTHS worth of sleep in their child's first two years."[6]

According to research carried out by the University of Warwick, new parents face almost six years of sleep deprivation following the birth of a child.[7] I read studies pointing out that when children sleep badly, there's a spillover to the rest of the family, increasing the likelihood of depression and lower overall family functioning. And a study from the University of California, Berkeley,[8] found that couples who reported poor sleep were much more likely to argue.[9]

Others have it worse still. Postpartum depression (PPD) affects between 7 and 13 percent of women after they give birth[10] and it's more common in single mothers, those whose partners and family are not supportive, mothers of sickly or premature babies and mothers lacking either financial resources or close family connections and friends.[11] Renowned child psychiatrist Bruce Perry writes in *Born to Love*[12] that PPD may be an adaptation developed to disconnect mothers from children who are born into circumstances where they're unlikely to survive, according to evolutionary theorists. So not getting too attached was a defense mechanism of sorts. Mothers were more likely to have children who survived if they saved their energy for existing older children or to reproduce again when food and emotional support were more readily available. By detaching, mothers could start to protect themselves from the pain of losing a baby—despite the fact that the detachment itself almost certainly made the baby even less likely to survive. Bleak, yes. But believable, considering the high child-hood mortality rates that characterize most of our evolutionary history.

For those of us who swerve the postpartum depression wrecking ball, there's still the monotony, madness and sheer mercurial joy of new roles and a brand-new life to contend with. I have entered a doorway through which there's no return. T goes back to work after paternity leave, and suddenly the buck stops with me. Even if I were raising a child with someone who wasn't obsessed with Lego and didn't lose his keys/wallet/phone on a daily basis, it's highly likely that there would be a primary caregiver, or a "managing parent." And that's me.

These new roles put our relationship under strain. We used to have fun. We used to go out. And eat dim sum. But how on earth did I think that because I had nice dinners with someone that we could raise a child together? Was I *high*? Was he? We are both doing something completely new and utterly stressful, on very little sleep. Parenting is unlike anything we've experienced before and with raising a child there's no finish line. Parenting is never "done." I wonder whether our relationship can take it.

The late psychiatrist Daniel Stern wrote in *The Motherhood Constellation*[13] that the shift in going from a couple to a couple + baby alters the mother's idea of her partner, "as husband, as father, and as man" (it's written in 1995, hence horribly heteronormative). New par-enthood is a time of conflict when all couples are a little messed up. Stern also points out that the terms used in popular culture to describe the impact of a new baby on a relationship are problematic. Babies are referred to as "marital glue," binding or even holding a couple together and suggesting that at least one parent would have left by now were it not for a gooey newborn "sticking" them together. The representa-tion of a baby as the "apple of a mother's eye" or even "the love of her life" ousts a father to the role of . . . what? "Provider? Protector? Enemy?" Stern suggests. Many men I speak to tell me that they worry about this—about not getting enough attention once a baby comes along.

"As though they're *also* babies?" a childfree-by-choice friend asks with some incredulity. I say nothing. Friend then proposes that "the other baby" could, perhaps, benefit from getting on with batch cooking or running the vacuum rather than engaging in an existential crisis, "but each to their own."

It's a fraught time for everyone so I'm willing to make allowances. But the point is, unless you get lucky and have an insanely relaxed baby along with two selfless, wise and capable parents, it's likely to be a shit-show. I am hypervigilant around the seven-month mark—the age that my sister died of SIDS. I am endlessly cautious about my baby's temperature, about bedding, about sleep positions and keeping him safe. T tries to be understanding—but how can he understand? It's hard. For both of us. And parenthood is the litmus test of any relationship.

Professor Paul Dolan, head of psychological and behavioral science at the London School of Economics, wrote a book on the delusion of *Happy Ever After*.[14] Dolan noted that while many of us consider marriage to be important, science shows that actually *being* married isn't much fun. Especially for women. He told an audience at the 2019 Hay Festival that: "The healthiest and happiest population subgroup is women who never married or had children."[15] Eli Finkel, a social psychologist at Northwestern University, argues that more of us than ever are dissatisfied with married life because our expectations of marriage have increased dramatically in recent decades. "[A] marriage that would have been acceptable to us in the 1950s is a disappointment to us today because of those high expectations," he told NPR in 2018.[16] A study as part of Mental Health Week 2019 found that life's so-called milestone moments—like love, marriage and kids—leave many of us feeling upset and down when the experiences inevitably fail to live up to our inflated hopes and social media hype.[17]

Love may be all we need at the start, but it takes grit to make it through. Starting a new relationship involves sex, suppressing flatulence, breakfast in bed and pretending we're cooler and cleverer than we really are. Being in a relationship involves Netflix, "I'd give that a minute" bathroom warnings, and emails about household chores. If there are children involved, two people who enjoyed each other's company morph into two people trying to raise a human being—on very little sleep and with far less disposable income than they might have had previously. Being cool or clever goes out the window in the trenches of parenthood and a good deal of each day is taken up with wiping butts, remembering to buy milk, then remembering to remember to buy milk because tiredness has made life ridiculous.

As the sole surviving child of a single mom who hasn't been massively successful in love, I have no working model of how a healthy long-term relationship works. What I have is a selection of snapshots: posed photographs in frames on our bedroom wall. There's one of T and I with our arms around each other, standing on a bridge somewhere, kissing. Another from our wedding day where we're both grinning. *We must have been happy then, right?* But then, everyone looks happy in wedding pictures. That's how weddings work: a picture is taken where everyone's smiling, it gets framed, and then years later, anyone looking at it assumes that they were having a lovely time in the good old days.

Take a look at anyone's walls or picture albums and we've all had a wonderful life, free from pain, suffering or bad hair days. Only we haven't, apparently. Because people around us start to get divorced. Some remarry, then split up, again. One obscenely handsome man I know tells me how he split up with his wife once kids came along "because it stopped being fun," but goes to great lengths to assure me he's still hopeful he'll remarry and "have more kids" and that things will "turn out differently next time." My first instinct is to burst out laughing. Then I have the sobering realization that, actually, what with his genetically gifted face and all, he probably *will* marry again. And procreate. And then leave when it "stops being fun." One friend tells me her theory: that no one should get married while they're in love. "While they're blinded by it," she says, "because it's only after you've got past the love part that you really see what someone is like." In the time of coronavirus, we saw just how many people's relationships depended on not having to spend quite so much time together. Because relationships are hard.

Remember the stats? It's estimated that around half of marriages end in divorce.[18] And yet, in a triumph of hope over experience, some people embark on the marriage merry-go-round a second, third or even fourth time (which, frankly, sounds exhausting). The data suggest that around 60 percent of second marriages end in divorce, and 70 percent of third marriages.[19] At which point, I feel as though I need to lie down.

The romance odds are clearly stacked against us and yet with unbridled optimism we plow on regardless. Until we don't. Until the "d" word starts getting bandied around. I look at friends who tell us they're getting divorced and think: How different is their relationship, really, from ours?

I'm not sure.

At the same time as this, my book comes out. *My* book. Full of *my* words. All 100,000 of them. These trundle out before being bound in a fancy jacket and transported to bookshops. And then it's out there, in the wild. I'm pretty confident that only my mother will read it, a) since T's a slow reader (if he's made it this far, congratulations!) and b) because who do I think I am, Danielle Steel? (I'm not, although I do love a statement earring.) I assume that my book will sink into obscurity along with 99.9 percent of tomes issued each year. But on publication day I get a call to say that the first print run has sold out and they're printing more. The second print run sells out a week later and the rights are sold around the world. I get reviewed. Most are good. One is bad. I remember this one vividly.

T's boss reads my book. His boss's boss reads it. I give interviews. My journalism training offered no advice on what to do when I'm on the other side of the spiral notebook (Don't fill the silence! Don't buy the proposition!) and I flounder, badly. TV crews want to come to my house. I am not strong enough or savvy enough to say "no" and so I let them. They are all, bar one, kind and gentle with me. But it's strange. Within a month I have two Japanese film crews and one Austrian crew in my home. As a terrible people pleaser, I allow myself to be directed too readily, in retrospect. I find myself variously chopping broccoli wrapped in a blanket as snow falls around me; pretending to read a book by candlelight as though electricity has not yet reached rural Denmark (it has); and having all my furniture rearranged for "just the right shot." I do a live TV interview with a sick bucket by my side during a stomach flu. I learn about "noddy" shots—where I'm shot from behind nodding obligingly. During one such interview in German (I do not speak German), I'm asked if we can "take that again" since I have been smiling while the interviewer was asking me about Denmark's occupation during the Second World War. "Oh! God. Sorry. Yes. Of course." Of course.

One morning, I bundle up my baby to drive to the scariest press interview yet and step outside to find my neighbor's teenage son leaning against my car, vomiting violently. He looks at me with drunken bloodshot eyes then raises a hand as if to say, "Stand by . . ." After a final expulsion, he stands up straight, wipes his mouth with the back of his hand and gives a nod of "all clear." I thank him, in bad Danish, and drive my car through his puddle of sick. I can't get to the radio studio on time so give the interview in a grocery store parking lot while my son defecates so ferociously that it leaks through his snowsuit and I become woozy with the smell of excrement. Getting what I wanted isn't quite as I'd expected.

I had thought that having a baby and a book meant I'd be immune to "sadness" or even "meh" days. I had thought that living in Denmark, the happiest country in the world, and researching happiness would mean that I would—via osmosis, perhaps—become a happy Dane with Viking DNA.[20] I had hoped that it would be rainbows and *Murder, She Wrote* reruns from here on in. But there's a strange sense of [*lowers voice*] anticlimax.

The Harvard University psychologist Dr. Tal Ben-Shahar knows all about this. He coined the term "arrival fallacy" in 2006 after experiencing its effects as a young elite squash player. "I had a dream of becoming a professional athlete, of winning championships," he tells me when we speak, "but the process itself was not enjoyable—the journey along the way came with physical pain and emotional struggles." Nevertheless, he was consoled by his conviction that once he "won," once he "made it," the pain would be over: "The reward would come and with it a new dawn will arrive!" But it never came. "After we win and we're *still* not happy we feel worse," he says, "because we're stripped of the illusion."

Ben-Shahar had to retire from playing squash professionally at age twenty-one due to injury and went on to study human behavior and thought patterns instead (squash's loss is our gain). "I came to realize that unrealistic expectations result in a sense of 'anticlimax,'" he says. "Every time."

So why don't we wise up to this?

"Conventional wisdom tells us that happiness is about the fulfillment of our goals," he says. "It's universal—we drink it with our mother's milk. We learn from a very early age that the path to happiness is 'success.' We inherit the idea that when we accomplish something, we'll finally be happy, as opposed to focusing on enjoying the progression toward the destination in mind." This is because "goal-hunting activities" stimulate the brain's reward centers and deliver a sense of accomplishment, says Ben-Shahar. We are all hardwired to enjoy the thrill of the chase. Dopamine rises as we anticipate a goal, then when we reach the goal, the dopamine drops off. We are biologically driven to pursue. But once we get the thing we want? We feel . . . nothing.

Achieving our goals is often less satisfying than we expect. And we're more susceptible to anticlimax or arrival fallacy if the goals we pursue are "external," says Ben-Shahar. So chasing money, power, public approval or parental validation are all likely to end in disappointment.

Internal motivation in the form of an *intrinsic* goal is a smarter approach. "This means it's something we want based on our values," says Ben-Shahar, "something meaningful that we care about. Not just what our parents [value] or society values." So exercising to feel good would be an example of an intrinsic value, while exercising to look hotter and impress people would be extrinsic. Working because we enjoy learning or are passionate about what we do = intrinsic. Working because we want good grades or to make a lot of money and gain prestige = extrinsic, and so on.

But even with intrinsic goals, we need to manage our expectations. "Studies show that the higher your expectations, the lower your happiness or self-esteem," he says. "If you have unrealistic expectations you will be disappointed more often. But many of us are sustained by the illusion of 'success' to pursue unrealistic goals," says Ben-Shahar.

What, like parenthood? I ask, skeptical. Can it really be that one of the most "natural things in the world" (allegedly) could qualify as an unrealistic goal?

"Yes," is the answer, swiftly followed by "especially these days." "Having a child is the ultimate unrealistic goal today," Ben-Shahar says. "Kids for most people are incredibly meaningful. There are many moments of pleasure but there's also frustration and anxiety and panic.

There are many moments of unrealistic expectations when it comes to becoming a parent."

And writing books?

"Ditto."

And marriage?

"Same."

Oh.

I feel I should carry a laminated card in my wallet that reads: "My name is Helen and I have arrival fallacy due to unrealistic expectations of marriage, motherhood and making a living as a writer."

Once I learn the term arrival fallacy, I start seeing it everywhere. In friends, in family members, in obscenely handsome divorcees, even in the dreams we're sold on social media and TV. Things I regularly tell myself include:

Once I do [insert thing here], I'll be on track.

After this busy week/month/year, I won't have to work so much.

When I finish this, I can spend time doing things I enjoy.

Soon, I'll take my foot off the accelerator . . .

But I never do. I believe, erroneously, that reaching my "goals" will make me feel amazing and that any agonies along the way are unavoidable collateral. Because the "goal" will be "worth it." Because: "goal!" Right? Wrong.

Numerous studies show that we consistently overestimate the amount of joy something will bring us and misjudge how events will make us feel. A 2007 article for *Harvard Business Review* even described how overachievers can become addicted to the adrenaline that comes from continually challenging themselves, naming this "summit syndrome."[21] I know a few adrenaline junkies—men and women who crave the highs of extreme physical challenges. Some even make a living out of it.

Like Ben.

It's time to meet Ben.

10

Summit Syndrome

B EN SAUNDERS IS one of the world's leading polar explorers who's covered more than 4,350 miles on foot in the polar regions. He led the Scott Expedition, the longest human-powered polar journey in history, and the first completion of the expedition that defeated Captain Scott and Sir Ernest Shackleton. He's also a friend of my friend Tony (Tony of "putting me on the dating site where I met T" fame. Keep up . . .).

I first met Saunders back in 2004 when he was preparing to ski solo to the North Pole and I was preparing to eat frozen candy for dinner (see Chapter 4). It was a time when Nickelback was in the charts and I'd just had to explain to my mother why Kelis's milkshakes were bringing all the boys to the yard, but I remember Saunders telling me that he'd been listening to a lot of Whitesnake lately.

Whitesnake?

"Whitesnake." He nodded. David Coverdale's rock classic "Here I Go Again" proved itself to be a fantastically fitting anthem to get him geed up before braving the ice, solo. I hold him personally responsible for my love of dad rock ever since and the fact that Van Halen's "Jump" has been my "break in case of emergency" song for the past fifteen years.

Saunders became the youngest person ever to ski solo to the North Pole on May 11, 2004, and was asked to give a TED Talk on his achievement (back when no mere mortals got to do TED Talks). On paper, he was winning at life. But it didn't feel that way.

"Reaching the North Pole was the ultimate anticlimax," he tells me when we speak again in the present day, "not least because there's nothing there. Not even a pole."

What? No pole?! Winnie the Pooh lied to us?

He nods. "Actually, I knew there wouldn't be anything there, intellectually at least; with the sea ice forever shifting, there could never be anything permanent on the North Pole. But I still expected . . . something."

Like what?

"I expected to feel different."

Ah yes, that.

Instead, he felt strangely numb. After looking around at the bleak expanse of nothingness, he sat down on his sled, pulled out his satellite phone and dialed three numbers: his mom, his girlfriend and his sponsor.

"I got three answering machines." Ouch. He came home, still hopeful of a hero's welcome. But at the airport there weren't crowds of people or press: just his mom and his brother.

"This trip had been my goal—my mission. I was expecting an overnight feeling of contentment," he says. "So when it didn't happen, I had quite a low."

He tells me he can identify strongly with the idea of arrival fallacy.

Ever the optimist, Saunders set out on another record-breaking expedition in 2013, this time to retrace Captain Scott's ill-fated journey to the South Pole on the longest ever polar journey on foot.

"It was groundbreaking: this was the journey that defined Shackleton. The journey that killed Scott. It was one of the greatest challenges known to humankind. And we made it!"

Hurrah! And then what?

"Well . . . then, for somewhere that had been so important to me for so much of my life—the South Pole—I was surprised by how *shoddy* the actual Pole looked." A hundred and one years previously, Captain Scott had stood on the same spot and written in his diary: "Great God! This is an awful place."

"So the clues were there," Saunders admits now. "But still, there was a sense of disappointment. Completing this journey had become an obsession that consumed much of my adult life." He waited for that sense of euphoria—his *Rocky* moment—but it never came. "In fact, life carried on much as normal," he says.

With admirable honesty and humility, he admits to searching news reports and honors lists for his name, "for some public acclaim." But all in vain: "My ego was hoping for more public recognition of what I'd done, but it wasn't there. All my drive, and energy, and all those years . . ." He tails off before clarifying: "I had thought that through ambition, application and persistence I would one day find happiness, inner peace and validation. Only it didn't come. Instead, after every trip, I'd get these post-expedition blues."

I tell him that he "wins" at arrival fallacy and summit syndrome. But his motivations to achieve stem from a similar place to my own. I know a little about his childhood—not least from Tony, who likes a chat. But now, he tells me: "My parents divorced when I was young and my father vanished from my life when I was eleven. I joined a long list of driven men who have this missing father figure, from Lance Armstrong to the explorer Ranulph Fiennes and an alarming percentage of politicians." He now admits that a lot of his early drive was tied up with "wanting to show off" in an attempt to prove himself to his absent father. "I can't remember any conscious sense of loss, although looking back now, it's clear that I was looking for a model of 'how to be a man.' I felt a need to prove myself. A kind of, 'Dad! Look at me!'"

Malcolm Gladwell points out in his book *David and Goliath*[1] that the loss of a parent frequently propels people into adult life with a higher than average level of ambition. So-called eminent orphans are defined as those who experience the loss of a parent before their eighteenth birthday. Gladwell wasn't the first to notice this—the US clinical psychologist J.M. Eisenstadt linked parental loss and genius back in 1978.[2] Many celebrated figures grew up without either of their parents: from Malcom X to Marilyn Monroe, Steve Jobs to Jamie Foxx, and Andy McNab to Aristotle. Even more grew up without mothers (Tina Turner, Madonna, Bono, Eleanor Roosevelt, Marie Curie, René Descartes, Queen Elizabeth I—though admittedly this was because her dad, Henry VIII, had her beheaded). Politicians are particularly prevalent in the eminent orphan category and almost a third of US presidents lost their fathers while they were young (George Washington, Thomas Jefferson, James Monroe, Andrew Jackson, Andrew Johnson, Rutherford Hayes, James Garfield, Grover Cleveland, Herbert Hoover, Gerald Ford and Bill Clinton). Turns out many of us grow up with daddy issues.

Saunders was reunited with his father in his late thirties and discovered that his dad had kept a box of newspaper cuttings about his son, proudly charting every reported achievement.

When I hear this, my nose twitches and I suddenly develop brimming eyes.

I wish my dad had done this.

Immediately I berate myself for thinking about him. For hoping. I'm annoyed with myself for still letting the loss of him get to me. Must. Try. Harder. But driven by loss and rejection, we strive. It's not healthy, but feelings of abandonment certainly get us off our asses and spur us on to *do* something.

Could it be that we need pain to pursue our goals?

"No," says Ben-Shahar when I ask (he doesn't beat around the bush). "You can be successful by being an optimalist rather than a perfectionist. And you're less likely to be unhappy this way, too." This is a message that may be difficult to hear if we've spent years pursuing success at the expense of peace and contentment. The idea that we could have been successful *and* less stressed might stick in the craw of anyone who's been busting a gut for years. "But I prefer to think of this as 'sunk cost,'" says Ben-Shahar. "If you've dedicated your life to perfection, it's *still* time to cut your losses. We all make mistakes. [*mic drop*] It's about learning to be good enough."

Ah yes: "Good enough." About that . . .

The British pediatrician and psychoanalyst Donald Winnicott came up with the phrase "the good enough mother" in 1953 after studying thousands of mothers and babies and observing that children *benefited* when their mothers failed them in manageable ways. The "ordinary devoted mother" turned out to be far better than the mother who had unrealistically lofty ideals about how to parent. "He was talking about mothers but you can apply the 'good enough' to life, to careers, to relationships, too," says Ben-Shahar. "In fact I'd go so far as to say 'you *should*.'"

He's the second person to recommend I "Winnicott-it" this week. The first was the neuroscientist Marwa Azab whom I get in contact with after watching her TED Talk about highly sensitive people [*waves*]. During a late-night chat, we end up talking about arrival fallacy. After a degree in psychology and completing masters coursework in research psychology, Azab was midway through her PhD in neuroscience when her third child came along. "It was hard. Really hard: three kids and a PhD? I spent *many* days crying, but my desire to get a PhD pushed me through."

And she got it—"on December 2, 2012!"—but then experienced a minor depression afterward. "I kept thinking, What next? I'd worked really hard to reach this goal," says Azab, "and I had arrival fallacy, 100 percent. From a neuroscientific point of view, this makes perfect sense. But knowing what I know, there's also a sense of 'what do I do about it?'" This can be hard on the ego, Azab says, "especially as a woman and especially in science. I have to be 'good enough' at more things than some of my male colleagues."

Such as?

"I want to be a good parent, a good partner, to be reliable in my career, but I have to say 'no' quite a lot," she says. "I'm offered opportunities that I have to say 'no' to, because I also have a family. I have commitments. I know lots of men who just can work late, or go on trips, or take opportunities. But I don't have that luxury. I work late, I work early, and this comes at a cost. But I put the work in. It may not be the best that I can do but I get the job done. I have to think about my baseline, and as long as that is still acceptable then that's okay. It has to be okay."

To stay "constructively sad" rather than "besieged by self-loathing sad," we have to keep on the "good enough" track. Ben-Shahar recommends ensuring that the thing we *think* we want is really the thing we want. "You need to make sure the goal you're pursuing is truly meaningful—not just an idea of something you want," he tells me. "It should be an intrinsic or self-concordant goal—the kind we pursue out of deep personal conviction or a strong interest—rather than anything external or extrinsic."

Next, we need to enjoy the ride.

"Take pleasure in the progression toward goal achievement," says Ben-Shahar. "This is known as the pre-goal-attainment positive effect, when you focus on the process and enjoy it, you take pleasure from the atmosphere of growth and the present moment." He tells me about a bracelet he wears to remind him to be present—a bracelet with the letter N on it for "NOW!"

"I also have my phone set up with a calendar notification to remind me to meditate daily."

Urgh: meditation. I am not good at meditation. I get too easily distracted.

"Do you actually *like* meditation?" I ask, skeptical.

"I need it," he says.

And yet you still need reminders?

"We all need reminders—and it's an ongoing process. Think of an athlete: she doesn't stop practicing just because she can run the race now." Fair point. Though if a Harvard psychologist still needs reminders, it'll take work to turn the steamroller around for the rest of us.

Saunders has made a start. "Slowly, I began to 'get it'—that reaching goals wasn't everything," he says. I knew it was up to me to decide what was it going to take for me to reach that point of contentment. It's taken a while, but now I'm . . . happy." I believe him: he looks it. "I've realized that pinning my self-worth on external validation is futile, because the more we accomplish, the more people we encounter whose achievements can make ours feel insignificant."

I nod: I *get* this.

There is always someone doing better than us. Envy is an ugly emotion. But is it avoidable? I put this to Ben-Shahar, who puts me in my place: "There are only two kinds of people who don't experience negative emotions like guilt or envy: the first are psychopaths, and the second type are dead."

I push it: even you?

"I am neither dead nor a psychopath, so of course sometimes I feel these things. The difference is that now I can be an observer. I can notice myself feeling a certain way and think, Oh that's interesting. I will accept the fact—though it's still not necessarily welcome. But the older I get, the more acceptance I have, the more permission I give myself to just be."

Just. Be.

So easy to type. So tricky to implement.

It's the least sexy piece of advice I have been given but I try it. I work on acceptance. I stop reading reviews. I come off social media for a while. I try to be present for my son. I play and read to him for hours each day. Until, eventually, he stops crying (so much). And then I work on my marriage.

The relationship expert Dr. John Gottman found that happier couples exchange at least five times as many positive statements as negative ones.[3] So I try being nicer. Psychologists have also found that when couples begin distancing themselves, one cardinal sin is to forget their history together and *why* they chose each other in the first place. Dr. James J. Ponzetti from the University of British Columbia studied 124 spouses and discovered that those who could highlight the basis for their marriage and all the positive reasons for getting hitched felt their marriage to be justified and affirmed.[4] Recollections of how it all began and regular reminiscing can help counteract the daily resentments and rough realities of long-term relationships—from the pressures of children, careers, money worries and general irritation to the inability to get dirty socks in the laundry basket (just for example . . .). As far as I can recall, T and I really did bond over a love of dim sum, and possibly otters, which, looking at it in the cold almost-light of a November day, seems no basis at all for a marriage.

We grow up thinking that falling in love is *it*. Then when we've reached *it*, there's a tendency to think that kids should come next. Many of us are raised to view parenthood as pivotal for the development and maintenance of happiness in adulthood, as Bibi Lynch observed. We think that the emotional rewards of having children will outweigh the emotional and financial costs associated and that a child will "complete us." But the reality is that it's hard.

That's okay, I think: I can do "hard." I chose this. I wanted it, desperately. I still do. But I'm not going to soft-pedal parenting or misrepresent the experience the way it was misrepresented to me. So there.

Research unequivocally shows that life satisfaction dips when there's a baby in the house[5] (especially when they're power-blasting walls with puke) and sociologists from Wake Forest University found that parents were more depressed than nonparents, no matter what their circumstances.[6] In one Princeton study, parents rated looking after their children as being "about as enjoyable as doing housework."[7] It's not all bad: researchers at Heidelberg University in Germany[8] found that parents were slightly happier than nonparents in retirement, but only once their kids have left home. To which T responds: "That's not happiness: that's *relief*." We probably have eighteen years of this ahead of us. I love being a parent. I'm grateful for my son and would gladly lay down my life for him/kill a medium-size woodland creature with my bare hands. And children are pretty funny, so there's that. But still, parenting = tough. And there's no guarantee that we or the person we chose to do it with will have a natural flair for it, either.

Three friends end up with individuals who are terrific parents but terrible partners. Others turn out to be pretty good at the partner part but awful with children. We might get together with someone because we admire their childlike qualities. But these aren't quite so appealing when we're trying to run a not-for-profit daycare center with them. We can yoke ourselves to someone who's faultlessly pleasant but fiscally incontinent. Or careful with money but ambivalent toward the idea of chores. In the couple set-ups I see around me, most partners (of either sex) are far from covered in glory. As one single friend puts it: "I don't know why everyone puts their head on one side and asks if I'm okay, like I won second prize in life because I'm not in a relationship—I haven't won second prize: I've won a different first prize. Relationships look rough!" She's right. They are.

Our baby reaches his first birthday. There are balloons and cake followed by an argument about whose turn it is to empty the diaper pail (it's definitely T's). One day, in a coffee shop, we are having a "difference of opinion" about something, crackling with energy, when I hear our "heated discussion" being relayed in Danish by the party behind us. *Oh great: now our fights are being translated.*

Things have got so bad that we pass for "entertainment."

Eventually, T and I go for couples counseling. This may have been helpful, were it not for the fact that the therapist insists on calling my husband "James." His name is not James. His name is nothing remotely like "James." We try to gently correct her: I refer to him repeatedly as "not-James." He tells anecdotes where he refers to himself in the third person as "not-James." I even email, signing off as "Helen and not-James." But to no avail. Every week, it's the same: "So, James, how have you been?"

James-gate becomes so ludicrous that shoulders start to shake and we have to squeeze each other's hands to avoid losing control. This has a strangely bonding effect. We keep hold of each other's hands after we've said goodbye to our therapist. The same happens the following week. After a few months of marriage counseling, we are so united by James-gate that we decide to stop. I'm not sure it's supposed to work like this, but "James" and I start getting on better. We sleep better. We sleep with each other. Maybe this was the therapist's cunning plan all along. Perhaps she's a maverick who doesn't play by the rules but gets the job done. Like Axel Foley from *Beverly Hills Cop* disguised as a middle-aged white woman in a sensible sweater. Who can say?

Neither of us has changed particularly. And we still aren't immune to arrival fallacy or summit syndrome—though we are at least *aware* of it now. I can still hide in my work when real life feels risky, and my husband still pursues a shopper's high whenever the opportunity presents itself. He is becoming a good father though, I see now. Someone who wants to give his child everything. Including a sibling.

The strange itching of broodiness—a kind of madness—begins to come over him, first, this time. I think I'm "safe" until I hear a baby crying in a store and find I'm crying, too. After that, the pangs start up again. The months that pass do not ease this: I want another child. On paper this isn't wise: we live far from home, it was a long hard road to conceive the first time around and we have only just become friends again after James-gate. We're not even sure it will be possible. But hormones have a lot to answer for. We become greedy. We push our luck. Rest and relaxation don't cut it this time—it's impossible to rest or relax with a toddler in tow and there are no more "miracles." So we end up sitting in a fertility clinic. Again. Our refrigerator at home gets stacked with boxes of fertility drugs. Again. But this time around, IVF works.

It will be fine, we tell ourselves, gluttons for punishment that we are. We have no blinders this time: sure, babies are hard, but we know what we're in for! We've done it once already! We've got all the gear! We have no idea.

"There's one," says the doctor as she tilts the monitor toward me during my first scan, "and here's the other one."

It's twins.

I'm having twins.

11

Get Some Perspective

WHILE BEYONCÉ ANNOUNCED her twin pregnancy to the world with a lingerie-clad Instagram post that broke the internet, I announce mine by puking into a trash can at my son's preschool.

Responses are mixed. "Congratulations" certainly makes the top ten, but it's beaten out by "How will you manage?" "You'll have a lot on your plate!" and "Oh God, I could not cope with twins . . ."

I'm not sure I can cope, either. But then a dad at the school gates tells me that he grew up as one of nine "and we were all homeschooled."

How? I reel.

"With patience," he tells me.

I'll bet.

"Also, our parents kept us streamlined: we were each allowed an interest—a *thing*. That could be a hobby, a sport, an instrument or 'friends.' But you had to pick one. And stick to it." Right. Good.

A mother of triplets meets the news with a combative stare that says, "Don't even *think* about whining to me: woman-up!" So I try. My body tries. But it's growing what turn out to be two normal-size babies and two placentas, simultaneously, which, we can all agree, is *not* normal.

I don't mean to compare myself to Beyoncé (this is a lie: it's essentially my life goal) but she and I develop along similar lines during our two-for-one pregnancies. From two distinct clusters of cells, I begin to swell like Violet Beauregarde in *Willy Wonka & the Chocolate Factory*. I get rounder by the week and once, a kindly neighbor informs me, I expand "a few good inches" over the course of an afternoon. Beyoncé and I both put on almost 60 pounds during our twin pregnancies, only she's a statuesque five-foot-seven Amazonian goddess and I'm a five-foot-three Hobbit who this time ten years ago was recovering from anorexia.

I struggle to breathe, metaphorically and physically, since there's less room for my lungs than usual. The heart works 50 percent harder during pregnancy and I am hot, All The Time. I can no longer sleep lying down or sit at a ninety-degree angle, because my heft continues past the point where my body would normally hinge. This makes the digestive process difficult. Both ends.* My pelvis is now jelly, so a wheelchair is offered to shuttle me to and from weekly check-ups for my "high-risk geriatric pregnancy" and a grab rail is installed in the shower. Walking feels as though knives are slicing my abdomen and no one's sure the three of us will make it, so I'm put on bed rest, with two months to go.

Propped up by several dozen pillows to a forty-five-degree angle— the semi-sweet spot between "squashed babies" and "squashed internal organs"—I lie there. All day. Every day. I develop bedsores on my hips and tailbone—ulcers on the skin and underlying tissue resulting from prolonged pressure. I have to be helped in and out of bed and turned at intervals, like a pork chop on a grill. All I can do for myself at this

---

* If you can't sit on the toilet, it's pretty tricky to go to the bathroom . . . as I discover.

point is think and feel and remember and . . . worry. Will my babies make it? Will I make it? What will happen if T has to look after our son by himself? What will happen if we all make it through in body, but not in mind? I've read "The Yellow Wallpaper,"[1] I know the drill.

My world shrinks to four walls. Friends visit. But most of the time, it's just me. I know myself well enough by now to be aware that I'm teetering on the edge. So I write, until I can't write anymore, because the pain is so bad and I am tired and the words start coming out like this: *kj8f7g****%oq9rjw/;fu'yfw.f* (too postmodern for today's literary tastes).

I remind myself that this is temporary, that in two months I'll have two babies. I remind myself that I am ridiculously fortunate, in so (so) many ways—and that in the history of the world, this is nothing. I remind myself that things have been far worse, for many, forever. But "me" doesn't quite believe "me" ("me" is a skeptic like that). So I prove it by immersing myself in a short history of sadness. This is a) less miserable than it sounds and b) excellent for putting things into perspective. Also, c) should come with the caveat that, as we discovered in Chapter 6, the "history of the world" as taught to someone raised in the 1980s and 1990s is far from exhaustive. So I start over.

Early Egyptian, Chinese and Babylonian civilizations, I learn, viewed sadness as a form of demonic possession and used corporal punishment and starvation to "drive out" the demons. So, you know, at least I'm not being told that I'm possessed by the devil. The idea of the "wandering womb" also emerged in ancient Egypt, later to be termed "hysteria" from *hystera*, the Greek word for uterus, and was thought to explain all forms of emotional excess in anyone possessed of a uterus. Wombs, they believed, went "wandering" about the body, clogging things up and causing all manner of ailments and unwanted emotions. Things weren't much better in ancient Greek and Roman times, although doctors sussed that low mood might be both a biological and psychological issue and you were likely to be prescribed an Insta-friendly regime of gymnastics, massage, special diets and regular bathing to alleviate symptoms. *Melancholia* emerges as an "illness" in the writings of Hippocrates, thought to be brought on by an imbalance of the bodily "humors" or fluids. Hippocrates believed that our bodies were composed of four substances: blood, yellow bile, black bile, and—wait for it—*phlegm*. Any sickness or disease in the body was the result of an excess of one of these fluids, and the doctor's job was to bring the humors back into balance by purging or bloodletting, usually (it's fair to say that Hippocrates had never spent much time with children who are, in my experience, 80 percent phlegm but remain remarkably upbeat).

By the Middle Ages, feeling sad essentially meant that God hated you. For clerics in medieval Europe, melancholy was a sign that you were living in sin and in need of repentance. Ten Hail Marys all round. In Chaucer's *Canterbury Tales*, written in the fourteenth century, despair and apathy were linked to slothfulness—one of the seven deadly sins. Being in possession of "outrageous sorrow" or a low mood meant that you couldn't perform your quota of good Christian deeds and so would probably end up in purgatory (thanks, Chaucer).

Renaissance writers and philosophers went a little lighter on sadness, adopting the idea that sadness was linked to creativity and getting hot and heavy on the idea of the "tortured artist." Being sad also meant that you were considered *closer* to God in the minds of many a Renaissance man (let's face it, no one was listening to women back then). In 1590, the writer Edmund Spenser even went so far as to endorse sadness as a marker of spiritual commitment. This strikes me as a convenient form of doublethink since British involvement in the slave trade also took hold in the sixteenth century—a trade the Brits would come to dominate as the leading trader in human lives across the Atlantic—so "sadness" was hardly in short supply. In stark contrast to Chaucer's theory, by the sixteenth century there was now an idea that if you were "happy," it was likely to be because you were getting your kicks from something not altogether holy—like sex or alcohol. This is unfortunate when you consider that a gallon of beer was considered a daily staple—like bread—in the 1590s. Each man in the English navy was furnished with *eight pints* of the stuff per day—so intoxication was commonplace.[2]

Unfortunately, my high school knowledge of medieval history grinds to a halt at this point and dusty old textbooks and household reading only take me up to the 1600s. But getting some perspective has helped, and I want more of it. So to continue "having a word with myself," I have a word with someone else: Thomas Dixon, professor of history at the Centre for the History of the Emotions at Queen Mary University of London. Dixon has explored the history of tears and is an advocate of the importance of a historical perspective on emotions. "Life was so bleak for so many for so long—with things like infant mortality—that there's a lot we can learn about how to be sad from history," he tells me.

For example, in the seventeenth century, children, women and old men were considered more naturally prone to tears but crying in general was frowned upon. There was also a prevailing view that sweating, weeping and raining were *literally* the same things: instances of vapors, in the body or in the air, being converted into water. This meant any "eye sweat" that did seep out wasn't exactly your fault, but was a little unseemly and could probably be avoided with better bodily discipline.

With advances in science and technology during the Enlightenment, thinkers began to consider how our bodies work from a mechanical point of view, seeing sadness as a "malfunction" of the human machine. But then eighteenth-century physician George Cheyne came up with the theory that "melancholy" was caused by all the newly acquired comforts and luxuries made possible by mechanization. Not enough toiling the land: too much sitting around *thinking*. To counteract the effects of such formidable decadence, Cheyne prescribed vegetarianism (a regime he apparently found it impossible to adhere to personally).

Then came the French Revolution in 1789 and, as the academic Rachel Hewitt puts it in *A Revolution of Feeling*,[3] "the decade that forged the modern mind." She hypothesizes that emotions as we know them today are a direct consequence of the French getting mad about cake/poverty (I paraphrase). "The French Revolution is described as 'sentimental humanism,'" explains Dixon, "this new 'cult of feeling' that quickly descended into 'brutality.' There was a sense that 'if you *feel* too much, look what happens!'"

From the nineteenth century onward, imperial powers were highly racialized. Crying and "feeling too much" was considered "other" and lower while Europeans—who supposedly didn't "feel" quite so much—were somehow "superior." "Tears at this time were thought of like an appendix," Dixon explains, for both sexes, "a useless remnant from our past." An unnecessary annoyance to men and a regrettable inevitability in "the weaker sex": women. But there was plenty to cry about with infant mortality rates still high and life hard for so many.

Many dealt with such short, brutal lives by telling themselves that it was "God's plan" and so accepting their lot—in the great tradition of Christian suffering. But people in the 1800s and 1900s also had rituals—symbolic activities performed to alleviate grief. Researchers have proven that rituals can help us come to terms with grief, following the loss of anything from a job to a relationship or a loved one. A 2014 study from Harvard University found that grieving rituals are crucial for regaining a sense of control[4]—something many will have experienced firsthand during COVID-19 when the ban on public gatherings made funerals impossible.

The First World War (death toll: between 17.5 and 40 million) and then the influenza epidemic of 1918 (50 million) also put a temporary pause on extravagant funerals and rituals, with the sheer scale of the loss of life from these atrocities making such mourning impossible.

The First World War devastated the lives of a generation, but the trauma of war didn't end on the battlefield. Tens of thousands of men on both sides returned from the trenches reeling from the horror of war: blind, deaf, mute or paralyzed; with acute insomnia, unrelenting anxiety, facial tics or chronic abdominal cramps. Doctors couldn't find any physical damage to explain away the symptoms, so the term "shell shock" was coined in 1915 in the medical journal *The Lancet* to describe the trauma soldiers had undergone. Four-fifths of shell shock cases were never able to return to military duty and were instead forced to grapple with their trauma in silence, since admitting "weakness" was considered "unmanly." Grief and sadness were taboo once more. And another stalwart of British culture was ready to step up and dictate the way many characters of influence coped (or rather didn't cope) with sadness: public school.

The novelist E.M. Forster wrote that these schools produced young men with "well-developed bodies, fairly developed minds, and underdeveloped hearts."[5] For the uninitiated, public schools are the not-at-all-public, fee-paying, private boarding schools that still exist in the UK today where children as young as six are sent. Though only a small proportion of the population go through this system, it impacts the entire nation since so many of the people running the country went to public school.

To put this into context, only around 7 percent of the UK population attend private school[6] but 60 percent of the UK's leaders have attended boarding schools.[7] This matters, not because these people are privileged, but because in many ways, they're *not*. The journalist George Monbiot calls boarding school a "form of abuse"[8] where boys and men, especially, have been taught not to show weakness or emotion (the two being viewed as synonymous). The psychotherapist Joy Schaverien coined the term "boarding school syndrome" in 2011[9] to identify a set of lasting psychological problems in adults who, as children, were sent away from home at an early age to boarding schools. Early boarding, she says, has similar effects to being taken into social or foster care, but with an added twist in that parents asked for it. *Paid for it*, even. Schaverien contends that premature separation from the family results in profound developmental damage, the repression of emotions in adulthood and even a form of PTSD. Many parents traditionally viewed this separation as a "cut and scar" technique that would make children "tougher" in the long run. But the consequences of spending one's formative years in "an expensive prison," as one former boarder describes it, are far-reaching.

I've had relationships with former boarders who, I can now reflect, suffered from all of the effects Schaverien described—occupying a different world to the one in which I was raised.[10]

Of course, not all boarding school experiences will be the same and not all boarders have husks for hearts. Henry Hitchings, the writer I spoke to about our habit of apologizing, is himself a former boarder who insists: "I'm not emotionally stunted as a result!" He assures me that: "Every school is different, of course—they have different cultures and values. But I was *never* told not to show emotion when I felt like it."

"Except in sports?" I check, thinking back to his comments about the importance of showing no weakness on the field.

"Except then," he admits. "But I think it has to do with the type of people who go to boarding school. They may come from families where they're encouraged to see expression of emotion as something to be frowned on. Some of the people I know who were unhappy at boarding school would have been very unhappy anyway."

I take his point. And yet . . . potentially uptight, emotionally repressed family + institutionalized, residential, same-sex (usually) education = years of therapy needed.

Back in 1882 at a boarding school in Ascot, an eight-year-old boy was whipped for damaging his headmaster's hat and for taking sugar from a pantry. "Flogging" was "a great feature of the curriculum," he wrote.[11] The boy's name was Winston Churchill.

A lot of blame for the repression of sadness has been lumped on Churchill. He's a far from unproblematic historical figure who believed in racial hierarchies and eugenics and was on the more "brutal and brutish end of the imperialist spectrum" as biographer Richard Toye describes him in *Churchill's Empire*.[12] Several psychologists describe him as a popularizer of emotional repression in his wartime speeches and rhetoric, suggesting that his impact on the culture of the UK and the US has been significant. To get through the Second World War, Winston Churchill urged allies to put on a brave face, keep calm and carry on (not his words, but indelibly associated with his spirit). Despite his own melancholy and dark moods, Churchill encouraged a buttoned-up-ness in the Western psyche that had a ripple effect out into society and still impacts us today—or so the narrative goes. But as Dixon reminds me: "Winston liked a weep."

On May 13, 1940, when Winston Churchill became leader, he told his Cabinet: "I have nothing to offer but blood, toil, tears and sweat." Churchill cried in private and in public. "When he authorized the destruction of a French fleet in Africa, for example, he cried," says Dixon, "and he would weep when he watched the movie *Lady Hamilton*." The story of Emma Hamilton mourning the death of Nelson was, apparently, a firm favorite with Churchill. "He watched this eight or nine times and would make everyone he was close to watch it as well," says Dixon. But then, crying over "art" has historically been far more acceptable.

Crying over art means that we're not crying for ourselves: we're crying for the plight of others. A far more "noble" endeavor that nevertheless offers relief, as we learned from the Tear Professor, Ad

Vingerhoets. We can appear to "hold it together" during much of our everyday life, then feel liberated to express our emotions when we're set off by artistic representations. This is neat: contained. Because then we can "stop the art" and get on with other things. There was a war on, after all, and "we" had to get through it, somehow. There were other, understandable motivations for locking up emotion. For a start, there were few outlets for one's woes and few people to listen to them, since everyone else was suffering, too. What's more, there was work to be done. And so a generation grew up judging that "getting on with things" was the best option—and that what wasn't spoken about couldn't hurt.

National pride became tied up with not showing emotion—in contrast with the "enemy forces." War correspondents reported on German and Japanese commanders "breaking down in tears" as they surrendered. But the allies were portrayed as "resilient."[13] Victory in WWII brought swift celebratory weeping, but then thin, broken men and women returned home—my grandfather among them, having spent years in a prisoner-of-war camp. But he never spoke of this and remained resolutely jovial for the rest of his life.

By the mid-twentieth century, advances in neuroscience gave psychiatrists and psychologists insights into how our brains actually work—rather than "guesswork 'n' demons." Scientists established that both chemicals and electricity make up our brain activity and that different parts of the brain are responsible for different behaviors and moods. So by the second half of the twentieth century we were at least allowed to be sad, but we probably shouldn't be doing it in public.

In the 1970s attitudes changed again. Baby boomers came of age as a generation more in touch with their emotions: softer, rebelling against emotionally repressed parents who had lived through war. "Feeling" was now okay again. But it wasn't without its problems.

"Boomers," born between 1946 and 1964, typically raised children to be more open, prioritizing self-esteem. "This was the beginnings of an emphasis on protecting the ego," explains the psychology professor Nathaniel Herr. "We started striving for 'happy' above all else," says Herr. We were allowed to "feel," but we'd better jolly well "feel" *happy*.

By the 1980s, psychologist Paul Ekman had identified sadness as one of our six basic human emotions—along with anger, fear, happiness, surprise and disgust (although this has now been established to be something we learn. See: kids + phlegm). So *being* sad was no longer a moral failure but letting children be sad was increasingly a #parentingfail.

The untimely death of Princess Diana was said to usher in a new, more feeling sensibility as people from all over the world came together to mourn someone who had meant so much to so many.[14] The 2000s brought with it a rash of reality TV shows that included—*depended on*, even—emotional backstories. Judges and contestants on shows like *American Idol* were now expected to show emotion and cry. "We all cry now," says Dixon. "Over what would have historically been viewed as 'very little.'"

I feel compelled to add the caveat "not everyone" here, following the Tear Professor's studies and my own research. Not *everyone* cries. Still. But if they did, that would be a good thing, wouldn't it? Or at least, better than the alternative of not feeling at all?

"Well . . ." Dixon hesitates. "Crying over everything is a massive luxury," he says finally. "The geopolitical situation could well change and we won't have this luxury." *Oh God*. It could also mean we've forgotten how to handle the big stuff over the last century. In the pursuit of happiness and with the loss of traditional "rituals" around sadness, we're in a worse place than ever before to identify, accept and experience the pain and suffering of normal life. At least, in much of the West. Because context matters. "Tears are produced by our beliefs about the world, so it's very important not to think that emotions are universal," says Dixon. "There's a lot of cultural relativism going on."

What we do isn't what everyone does, so for a better perspective on sadness and how to handle it, we need to go beyond our own borders and take off our culturally biased blinders.

Strap in . . .

12

Get More Perspective

I'VE SPENT YEARS researching happiness concepts around the world. I wrote about some of the unique cultural concepts around contentment in a book, *The Atlas of Happiness*.[1] But I could just as easily have written a book on global sadness and the ways that other cultures handle theirs. Because there's a lot we could be doing differently and plenty we can learn from the endlessly fascinating and enlightening rituals that help people around the world to be sad, better.

Take Greece—a country that's had its fair share of hurt in recent years but where a culture of expressing emotions rather than repressing them helps Greeks to stay afloat. So mourning in Greece is a big, public affair and Greeks traditionally believed that weeping together creates a bond between people. As one Greek friend likes to remind me: "When grief is shared it is halved." In Bhutan, crematoriums are centrally located, so children grow up with an understanding that loss and death are inevitable. Traditionalists in Spain might marvel at the body of the deceased behind a pane of glass and spend time with the deceased person, contemplating life and loss. For those electing to have a send-off Catalan-style, the body is laid out in a display case in the center of a room so that family can spend an entire day with it. There's a week-long period of mourning in Judaism with first-degree relatives "sitting shiva" to mark their respect. And Hindus mourn for thirteen days following a death, ending with a ceremony known as *sraddha* that's also performed on the anniversary of a loss.

Many cultures have conflicting or even contradictory rituals but researchers from Harvard University (notably Michael Norton, in work spanning decades) have found that it doesn't necessarily matter *what* we do—the important thing is that we're doing something. The 2014 study also found that we don't even necessarily need to believe in or endorse the efficacy of various rituals for them to help us mourn and regain a sense of control.[2] As the French sociologist Émile Durkheim put it in 1912: "Mourning is left behind, thanks to mourning itself."[3] Rituals—of any kind—help. But many cultures in the West now lack these.

We've never had a grief culture with as few rituals as we have now. But we need these collective rituals to help us express our emotions. And it's not just death that is grieved in a more helpful way in other countries. The Brazilians have a national day in honor of *saudade*—the Portuguese concept of melancholy and nostalgia for a happiness that once was, or even happiness we merely hoped for. Imagine that. A whole day dedicated to feeling sad about the fact that life hasn't worked out the way we might have liked. In South Korea, to help people appreciate life, some companies encourage employees to act out their own funerals.[4] Workers watch videos of people worse off than them—the terminally ill or victims of war—then write letters to their loved ones, before lying down in a wooden coffin to contemplate life and feel grateful for what they've got.

For Australia's Aboriginal and Torres Strait Islander peoples, emotions are traditionally framed and understood within the construction of spirit or *Kurunpa*—the life force and essence of Aboriginal life.

Kurunpa is vulnerable to the impact of trauma, grief and loss, social chaos, sorrow and despair;[5] and in common with much of world's more than 370 million indigenous peoples who've been mistreated, "sorrow" is in no short supply. Aboriginal and Torres Strait Islander peoples are twice as likely to be hospitalized for mental health disorders and die from suicide—with those aged fifteen to nineteen *five times* more likely to die by suicide.[6] But since 2019 a new culturally specific tool is used by doctors to screen for depression among Aboriginal and Torres Strait Islander Australians. It's called the aPHQ-9 and it's an adapted version of the existing PHQ-9 that includes modified questions such as: "Over the last two weeks, have you been feeling unhappy, depressed, really no good, that your spirit was sad?" It's hoped that this will remove any lingering stigma—especially among Aboriginal and Torres Strait Islander men—about mental health and speaking out about "negative emotions." And it's an incredibly powerful way of looking at the world (is your spirit sad right now?).

In Māori culture, the repression of emotions, good or bad, is the antithesis to the famous haka—a ceremonial group song with accompanying movements, characteristically involving stomping, shouting and powerful gestures—brought to a worldwide audience by New Zealand's rugby team, the All Blacks. But haka isn't about aggression. For Māori, strength and showing emotions are one and the same. The goal of haka is a reconnection of the body, mind and spirit. One Māori teacher I met described haka as "a way to orchestrate a type of unkempt energy a lot of people don't know they have, then give it back to them in a way they can understand." Traditionally, the practice had been limited to Māori communities, but today most Kiwis learn haka in school from an early age, regardless of whether they are Māori or not. When New Zealand mourned the victims of the 2019 Christchurch mosque shootings, the Māori Council commissioned a new haka taking a stand against hatred to commemorate the victims of the shooting (included in the Notes at the end of this book)[7] that was performed by everyone throughout the country in solidarity.

Grief is inescapable and the best thing we can do when we experience it is to come together and shore up our human connections, rather than becoming more divided or pretending nothing's wrong.

This is the thinking behind the South African concept of *ubuntu*, the belief in a universal human bond and the idea that "I am only because you are." Archbishop Desmond Tutu famously promoted this as a theological concept and now his granddaughter Nompumelelo Mungi Ngomane, author of *Everyday Ubuntu*,[8] continues his work and promotes *ubuntu* for a secular audience, too. "I feel as though sadness is resisted in many parts of the world, especially in the US," says Ngomane, who grew up in America but now divides her time between the US, South Africa and the UK. "It's most evident if you compare mourning in the US versus in South Africa," she tells me when we meet. "In the US, you have a funeral and then you're expected to return to work and your grief is considered 'done.'" But, as Tutu said: "suffering is not optional." It afflicts us all, so we have to learn how to handle it. "In South Africa, we're better at sitting with sadness," says Ngomane, "and we support each other. This is *ubuntu*." Empathy is key: "So we feel sad when the people around us feel sad. And we don't try to 'fix' sadness all the time, we can just feel it, too. Sometimes we need that."

Life won't always be fun, but some cultures have rituals and approaches to help us handle this. The Chinese *xingfu* is often translated as "happiness" in English but actually refers not to a good mood, but a good *life*—one that is sufficient and sustainable and has meaning. It isn't necessarily an easy, pleasant existence (in fact the Chinese character for *xing* represents torture); life may be hard, but it will have meaning.

The anthropologist Catherine Lutz studied the Ifaluk people in the

Western Pacific whose culture emphasizes nonaggression, cooperation and sharing.[9] Constraints of island life necessitate compassion and consideration for one's fellow human, best summed up in their unique concept of *fago*, meaning compassion, love and sadness—concurrently.

You can learn a lot about a culture from its language, and just as English is rich in terms for expressing embarrassment ("mortification," "shame," "discomfiture," "awkwardness"), the Welsh have more words to describe soul-churning love and the pain that comes with it. *Hwyl* refers to a strong, stirring feeling of emotion and fervor, while *hiraeth* is a raw, homesickness-type of emotion. In Poland, *Żal* refers to a combination of love, loss, sadness, sorrow, regret, hatred, melancholy *and* rage—"plus all the emotions in between" (according to my Polish publisher). It's also said to be indicative of Poland's troubled history. The Czech term *litost* is a state of torment created by the sudden sight of one's own misery—a kabuki mask of despair (a concept anyone who's ever hung around crying children will "get"; most only have to catch a glimpse of themselves in a reflective surface to intensify their howling tenfold). *Litost* is a word of such importance to the national character that the Czech writer Milan Kundera says: "I find it difficult to imagine how anyone can understand the human soul without it."[10]

And then there's Japan. I've spent time in Japan over the years for work, and the principles of *wabi sabi*—appreciating the beauty in imperfection and the transience of life—have been immeasurably helpful for me through recent tough times. But there's also *mono no aware*, or the "awareness of things," a term used to describe the pathos and empathy for all of life, its impermanence and the innate uncertainty of our existence. *Mono no aware* comprises both sorrow *and* a serene resignation and acceptance of the fact that everything and everyone must end—a knotty concept sorely lacking a descriptor in English.

Psychologist Jeanne Tsai, from the Stanford Culture and Emotion Lab, believes that we can learn a great deal from Eastern cultures, where experiencing both positive and negative feelings simultaneously is considered normal. She refers me to a study showing that when Japanese students experience success, they feel a mixture of emotions. "On the one hand, they feel happy that they succeeded, but, on the other hand, they also experience a fear of troubling other people," she says. "And that is related to how much they believe that they're responsible for the feelings of other people." As part of a communal society, there's a sense that we're all in it together and Japan is home to the increasingly popular phenomenon of *rui-katsu*, or "tear seeking," where groups of people gather to watch sad movies and have a good cry together[11] (they also have *ikemeso danshi*, or "handsome weeping boys" who are paid by women to wipe away their tears;[12] just saying . . .).

There are clear distinctions in the approaches to sadness in Eastern and Western cultures. But to truly get my head around sadness in a global context, I have to find out more about the country that straddles East and West: Russia.

It's no surprise that the country responsible for bringing the world Chekhov, Turgenev, Tolstoy, Gogol, Gorky, Nabokov, Dostoevsky and Pushkin is pretty good at "sad." Terms such as *toska* ("great spiritual anguish")[13] and *dusha naraspashku* (an "unbuttoned soul") pepper the language and "there's an idea that being sad makes you a *better person* in Russia," says Dr. Yulia Chentsova-Dutton, associate professor in the Department of Psychology at Georgetown University. Russian by birth, Chentsova-Dutton now explores how emotions emerge from the interaction of universal tendencies, cultural scripts and situational cues. We speak over Skype and she tells me, off the bat, that sadness is not only allowed in Russia—it's prized: "We believe that you develop through being sad." This is something she attributes to the Orthodox Church and the cultural models linking sadness and suffering with moral virtue: "There is a strong idea that because Christ suffered, suffering must bring you closer to God."

Even if you're not religious?

"Even then," she tells me, "the culture is so dominant that this view is largely ingrained."

Russian adults say that they "value sadness" in studies, "and parents will say that they want their children to experience sadness," says Chentsova-Dutton. "They endorse statements like 'sadness helps you connect to others' or 'sadness helps you appreciate the richness of life.'"

Russian parents willingly read their children "sad" books" to help with this and Chentsova-Dutton tells me about a few of the most famous stories that even very young children are reared with. In one, a little girl is crying because her ball fell in the river and floated away. In another, a poem, a toy bunny is left out in the rain and becomes sodden.[14] In a third, a toy bear falls out of a window. *Oh, how we laughed!*

"Children learn that the characters deserve compassion," says Chentsova-Dutton, "but the sadness of their plight is not relieved in any way. More than that: it is *relished*." She recounts another story about a little boy whose dad went off to fight in the Second World War. "At the end, the soldiers return from the war and there's a big party," she tells me. "But this boy's dad has died—so he doesn't come back.[15] The story ends with this real punch in the stomach. I was five when I first read that one."

I think of my son's bedtime reading. Everyone's always happy, in the end. But are we doing our children a disservice if we only expose them to upbeat endings? Are we failing to prepare them for the realities of life? Chentsova-Dutton thinks so.

"In the UK and the US, people have an almost adversarial relationship with sadness. If children's books show sadness, there is immediate repair. Whereas in Russia, you befriend sadness." She tells me about a time she wanted to impress a new teacher growing up: "I decided that the best way to do this would be to look somber. This would communicate that I was a good and thoughtful student. I already knew, aged eleven, that the very worst thing I could do was smile."

I've heard that Russians are "taught" not to smile from an early age. But can smiling really be so—pardon the pun—frowned upon?

"Yes," she says. She does not smile. "One of the most common disciplinary statements to Russian children is 'stop smiling.'" Because sadness and the reflection that comes with it are valued in Russia. "In America, if you ask people whether they would choose to feel 'happy' or 'sad' or 'nothing' ahead of a task, we've found that they'll respond 'happy' or 'nothing.'" But in Russia, people report that they *want* to feel sadness ahead of a task as they know it will help them with concentration. "These theories are hard to test, but many studies endorse a belief that sadness helps with concentration and is broadly good for social interactions." When we are sad, we are more relatable and even more *likeable*.

Chentsova-Dutton now has a daughter growing up in the US. "But I'm trying to institute a bit of Russian grit in her," she tells me. This can be difficult: "In the US there is such societal pressure against letting children experience discomfort, or be bored, or sad, or in pain." Chentsova-Dutton recalls a trip "home" when her daughter played in the forest with neighboring children only to be stung by nettles. "My daughter cried and ran home, distressed because her legs had been stung," she says, "but the Russian children were incredulous—their approach was: 'Of course your legs get stung, that is part of the adventure! You suffer! You feel pain and you feel sad and it helps you grow!'"

Sadness, I've learned, is a good thing in Russian culture. It can teach us something, even when it hurts. We shouldn't shield children from pain or downplay death, because this won't help them (us) handle loss when it inevitably comes calling. I thank Chentsova-Dutton and hang up, stiffened in my resolve.

Okay, I think, I can do this. Our pain helps us grow. Literally, in my case. Sadness has value. Purpose, even. So I focus my attention.

I incubate.

I wait.

I develop . . . what is this strange new sensation? It's something like . . . wait, can it be? *Patience?*

At thirty-eight weeks, my enormous babies are deemed "done" and I am wheeled into the hospital. A consultant wrenches out baby number one and I find I can breathe again. For the first time, properly, in months. I weep with relief. T faints and has to be removed from the operating room. But still: this is going to be okay, I think.

The second baby is pulled out with relative ease (relative).

And just like that, one section of my life is over and two new lives begin.

13

The Tipping Point

I T'S CHAOS. For a long time, life is chaos. T and I white-knuckle
it, with no slack: no room for anything to go wrong. Two babies
scream, the toddler screams. I pick one up to soothe them. The others
scream louder. And so on. If we work as a team, with military precision, everyone gets fed, clothed, washed occasionally, and we all get
through the day without picking up any communicable diseases. It's
fine ("*fine!*") because I love them all. But they don't necessarily love
each other. Not yet. The flame-haired three-year-old is instantly sus-
picious of these new arrivals and asks on several occasions when we
can "give them back." He is distraught when I tell him the answer is
"never."

A Singaporean friend tells me I have "lucky" twins: a boy and a
girl. And I feel lucky. Three and a half years ago, I thought I'd never
have children. Now there are three. Deep breath. There are moments
of sheer wonder when I look at them in awe: a "spirited" toddler and
two new little animals, so close that they sometimes suck each other's
thumbs. Then they wake up and it's chaos again.

The boy-twin is the spitting image of me as a child: cheeks you could gnaw on and thighs that come with rolls on them. Like a mini Michelin Man. My daughter is a slip of a thing: eyes as big as saucers, fair-haired, froggy. Familiar, too, in a way. It's my mother who finally points out the resemblance, the thing all of us have avoided saying: she looks like my sister. My mother finds the likeness emotional, understandably: looking into the face of a baby daughter who died. I find it emotional: looking into the face of my daughter and seeing my sister. The day she died is my first ever memory, despite the fact that I was only two and three-quarters at the time. It's strange and sad. What's stranger still is observing her alongside her sibling, who looks like me.

My mother refers to girl-twin as "my angel." This makes everyone uncomfortable ("as though she's died!") so I ask her to stop. She does and we all exhale slightly.

We allow our shoulders to lower fractionally further once both babies make it past the seven-month mark—the age my sister died. Another unspoken milestone is passed when they reach toddlerhood— the age I was when I lurched into cognition. I become acutely aware that they [*could*] remember anything that happens from here on in. But we are okay. Sure, we are bone-tired by the end of each day, but, emotionally at least, I feel like I can do this. My children are not my mother's children. And yet there's something strangely moving about watching mini-me toddle around, hand in hand with a mini-sister.

Then boy-twin is scheduled for a minor operation under general anesthetic. It's already been a difficult morning, denying one toddler food and drink ahead of surgery while trying to explain to the other that, for the first time ever, her twin won't be with her today, but somehow I peel my daughter off my leg and cajole my younger son into the car. At the hospital, he is prepped for surgery. I'm asked to wear what looks like a shower cap to enter the operating room and my son is immediately apprehensive.

"No hat, Mummy!" He wrinkles his nose.

"Mummy has to wear the hat—"

"No HAT!" He's more forceful this time. It isn't fashion advice: it's a command.

"Doctor says I have to wear—"

"NO *HAT*, MUMMY!"

Next he sees the mask that will be used to administer anesthetic. "NOT that!"

I try to explain.

He starts to cry. The gas is turned on. The doctors ask me to hold him still on my lap, clasping his hands by his sides to stop him pulling at the tubes.

"No smelly mask!" He's screaming now, fighting as I hold him close. An anesthetist and three nurses have to help restrain him.

"P'ease, Mummy!" ("l's are tricky when you're two.) The mask pumps out vapors. He looks at me in horror.

Why are you doing this to me?

"I'm sorry . . ."

After this, he stops asking for help.

Mom is no good: Mom can't protect me.

He uses the last of his fight to try to wriggle away: *I will run wild in this hospital if I have to . . .* The medics hold him. He stops struggling and his sobs sound further and further away, echoing down the pipe that is currently pumping anesthetic into him. The cries get fainter. Limbs stop thrashing. Until finally, he is limp. Thirty-five pounds of solid cherub in my arms (yes, he's heavy: me and him make for sturdy toddlers).

Four medics lift him off me and my heart stops.

"Can't I carry him? Can't I take him to the operating table? Can't I stay with him?"

"No," is the answer. "He's in safe hands."

I'm asked to leave but struggle to get my body to respond, paralyzed by gluey helplessness. I am aware, intellectually, that this is a low-risk operation for which my son needs to be under general anesthetic. But the animal, prehistoric part of me thinks I have just watched him die. Worse: that I've helped. Also: I have just watched a child who looks like me become lifeless and limp. My prehistoric brain is in bits.

I remember the actor Rory Kinnear once saying in an interview that playing Hamlet was so draining because, "the body doesn't *know* it's acting." My most basic self doesn't know about anesthesia. My basic self thinks that my child is gone.

As I wait for him to come round I think of how my mother must have felt. And then I stop, because it's too painful.

When my son wakes up, he is wary of me and of the world. He becomes afraid of going to sleep and wakes crying in the night—so the rest of us wake up, too.

Every night.

In the mornings, we get up, I make oatmeal. We find shoes, jackets and socks, making it through the day before brushing teeth (again!), pajamas, stories, bed. Repeat. Then T goes away for work. And it's just me.

I am woken up for the day by "I done a pee pee!" any time between 4:30 a.m. and 6 a.m. The smalls slurp a few mouthfuls of oatmeal, then I clean up the rest of the sweet slime from the table/floor/chairs/faces; select three sets of clothes; brush three pairs of teeth; and begin three sets of negotiations. "Yes, okay, you can take the spatula with you. And the Lego. No, it's your brother's turn with the monkey backpack. No, you can't take a hammer. Because your teacher said so. No, teddy doesn't 'love danger' . . ." I drop them off at day-care, work, eat at my desk, pick everyone up again before heading home to make food/run a craft emporium/clean up bodily fluids ("I needed to pee!" "Okay, coming! Wait, what do you mean, 'need*ed*'?" I fetch the mop).

Work is different now. "Can you cover a bomb going off in Sweden?" I'm asked one afternoon. "No! I've got a shepherd's pie in the oven," I find myself telling the editor on the foreign desk of a major newspaper. I work as hard as I did when I was employed by someone else, but for far less money. I read a study telling me helpfully that working mothers are officially 18 percent more stressed than other people[1] and that working mothers of two children are 40 percent more stressed. For single mothers or those of three-plus children, it's doubtless higher still. I don't find these results surprising, though I do feel instantly guilty for nodding along with them. For complaining. *I'm lucky*, I know it. I tell myself not to be so ungrateful. On loop. And I keep going.

Burn-out is included in the 11th edition of the *International Classification of Diseases (ICD-11)* as an occupational phenomenon[2] characterized by feelings of energy depletion or exhaustion; increased mental distance from one's job, or feelings of negativism or cynicism related to one's job; and reduced professional efficacy. The official definition of burn-out refers to stress in the workplace "and should not be applied to describe experiences in other areas of life." But parenting is a job. And if we're doing it on top of paid employment, the capacity for burn-out is immense. Only working mothers don't get the manly sounding "burn-out." We get "Rushing Woman's Syndrome."

I am asked to write an opinion piece on this newly diagnosed condition, coined by nutritional biochemist Dr. Libby Weaver who claims that the cult of "busy" is making modern women miserable, sick and hormonally imbalanced. I say "no" to writing the article, twice, because I am too busy rushing around to write about being too busy, rushing around. The editor perseveres. I query: "Why me?" There is muffled coughing down the phone. I get it: I write it.

I have a rare dinner out with some interesting people for work and then wake up the next morning in a funk at the seemingly

Sisyphean task of cleaning up encrusted oatmeal. I find myself wishing I'd never had interesting conversations about sustainability, politics and the economy because then I might not miss them so much.

I read a slew of books and articles about "time-saving life hacks!" or "leading a more centered existence!" while still being wildly successful in the hope that some of them might "save" me. These are all written by men who only mention in the acknowledgment section that "none of this would have been possible without the support of [*insert wife's name here*]" or that "this is what Daddy was doing all those weekends he wasn't around!" None of them are written by women with young children. Presumably because they're up to their eyeballs in encrusted oatmeal. My generation were promised we could "have it all," that we could have children *and* an intellectual life. So am I being greedy? Overly ambitious? Both? Possibly. And yet . . . I have very low domestic standards: I cook only the basics, my cleaning mantra is "how little can I get away with without attracting mice?" and I work around available childcare. But the task of keeping afloat seems Herculean.

The kids fight. The eldest still doesn't think much of his toddler siblings messing up his toys/room/life. Twin toddlers, used to sharing everything, find this unreasonable. Tiny, angry voices buzz. Growing up, I never heard a raised voice. My "family"—my mother and me, occasionally Orange Backpack—didn't shout: we simmered. Or displaced. But now there is a near constant din. Someone is always yelling. Or staging a dirty protest.

Another day passes where I may or may not have had time to shower. A sense of disquiet scratches like a pin and songs in a minor key make the space behind my eyes hurt. I am hollowed out. The daily irritations of traffic jams, electricity bills, tax and laundry start to make me feel a little fighty. I struggle to sleep at night, then wake up with each child in turn and stare at the ceiling for hours after every disturbance. I welcome in 3 a.m. far more often than I miss it. Four a.m. is similarly familiar. And by 5 a.m. I know that someone will wake up in the next hour, so there's no point hoping for a reprieve. By morning, I ache, unfolding my body and swinging legs out of bed to start the day again, drowning by inches.

Russell Foster is professor of circadian neuroscience at the University of Oxford and director of the Sleep and Circadian Neuroscience Institute. He told the BBC in an interview that even short-term sleep disruption is associated with profound brain dysfunction: "We lose our ability to lay down memories and then manipulate this information; to come up with innovative solutions to complex problems. We lose our empathy. We become overly impulsive so we do stupid and unreflective things." He adds: "The tired brain will remember negative stuff and forget the positive stuff."[3] This feels familiar. No wonder I feel bad.

"Sleep also clears away our brain's clutter," says the neuroscientist Dean Burnett, "the free radicals from neurological processes; all the molecular bits and bobs from the chemical breakdowns; all the stuff your brain needs to metabolize and all of the chemical by-products of brain functioning are cleared away in sleep. So if you don't sleep, that doesn't happen. The clutter builds up; clogs you up and then you can't function."

I can't function. And I start to feel very, very down.

My sleep deprivation becomes so severe that I begin to hallucinate. Colors at first. Then furniture, moving when it should not. Then cars. Changing lanes. When they have not. Hallucinations are normally experienced by people on hallucinogenic drugs (clue's in the name), those experiencing psychosis or suffering with schizophrenia or dementia. But it's surprisingly common for the chronically sleep-deprived, too.[4] The exact reasons why aren't known, but it's thought that the parts of the brain responsible for visual functioning get disrupted, or that dopamine levels are affected, or even that the brain becomes so tired it enters a strange new state of consciousness. To avoid becoming tired to the point of hallucination, we need to recognize the early signs of sleep deprivation—from a change in mood to impatience, increased irritability and difficulty concentrating. When these signs present themselves, it's time to make sleep a priority.[5] I try, but my life won't let me. And so, slowly, it becomes devoid of Technicolor. Like *The Wizard of Oz* in reverse. I'm stuck in Kansas. With no dog.

Not this, I think. Not this again. But the birds stop singing and one day the sun is so pale, I can hardly make it out at all.

T comes home and tells me he thinks I should go to the doctor. I am reluctant, for three reasons. First, I've been raised to be wary of "making a fuss." The second is that I live in Denmark but still don't speak Danish well and doctors tend to be rather judgy about this (a guilt trip I can do without at the moment). The third reason is that the only doctor who doesn't mind my bad Danish and so the one who has been officially assigned to my family is, regrettably, very handsome. And kind. And his eyes mist up slightly in empathy when he talks to me. And I have been half in love with him for the past four years. Consequently, I do not want to see dream doctor now.

T insists.

The imaginary cars won't stop changing lanes.

So I go.

I'm hoping that perhaps I have some hitherto undiagnosed tropical illness that a swift course of antibiotics will take care of. I'm hoping I can be "fixed" and that, under absolutely no circumstances, the way I'm feeling is a result of "life." I sit in the overly stuffy waiting room, my tongue heavy in my mouth, throat constricting, head in hands. Hot tears fall straight from my eyes onto the carpet tiles, making patterns of dark grey circles. And then my name is called.

Dream doctor looks at me as though he may weep. Then he takes me through the by now familiar questionnaire for depression. There are nine key symptoms and someone has to show five of these for a two-week period to "achieve" a diagnosis. But someone else could have four different symptoms with just one in common and we'd both come out with the same diagnosis. Weird, right? Here's why: in 1974 a young faculty member at Columbia University called Robert Spitzer was handed a spiral-bound manual of 150 pages (the *DSM-II*, last updated in 1962) and asked to come up with an updated diagnostic manual. This would become the *DSM-III*, used by millions of mental health practitioners in the US and around the world for decades to come. A daunting challenge, admittedly, but one that Spitzer took on with gusto. In 2007 the American psychiatrist Daniel J. Carlat, MD, interviewed Spitzer about his decision to stipulate five criteria as the minimum threshold for depression, to which Spitzer answered: "It was just consensus. We would ask clinicians and researchers, 'How many symptoms do you think patients ought to have before you would give them the diagnosis of depression?' and we came up with the arbitrary number of five."

Note: "*arbitrary*."

Carlat then asked Spitzer why he chose five and not four or six. He writes of his encounter with Spitzer: "He smiled impishly, looking me directly in the eyes. 'Because four just seemed like not enough. And six seemed like too much . . . there is no sharp dividing line where you can confidently say, This is the perfect number of symptoms to make a diagnosis.'"[6] Because every brain is different. And psychiatry is an imperfect science.

"We know almost nothing about the biology of the brain," admits Dr. Nelson Freimer, director of UCLA's Center for Neurobehavioral Genetics and a professor of psychiatry. "Really, there is just not that good evidence for a specific theory of depression."

Even with all the modern technology and skulduggery? I ask. Even with MRI scanners?

"Nope," he says, "they don't tell us anything conclusive. We still haven't much of a clue. I think anyone who says otherwise is much too optimistic." Practitioners can only work with what they have.

The *DSM-5*, updated in 2013, now includes three times as many "disorders" than it did in its first iteration in 1952 and is seven times longer. Given this expansion, it's little wonder that the National Institute of Mental Health now claims that one in four American adults suffer from a diagnosable mental disorder in a given year,[7] while the World Health Organization puts the global figure around the same.[8]

But remember: diagnosis is *arbitrary*.

Mental health organizations A Disorder for Everyone and Safely Held Spaces point out that there is little evidence that these very real and difficult experiences are best understood as "disorders" and "illnesses" caused primarily by chemical imbalances or genetics. They argue that these can instead be related to life events, such as trauma, loss, neglect and abuse, as well as wider social factors such as unemployment, discrimination, poverty and inequality. Dr. Lucy Johnstone, a campaigner for the de-pathologizing of sadness says: "The key question when encountering someone with mental or emotional distress shouldn't be, 'What's wrong with you?' but rather, 'What's happened to you?'" By making "what's happened to us" into a psychiatric condition, we're denying that it's a normal part of life.

And we've all had "stuff" happen to us.

Dream doctor tells me that I *could* be depressed ("You certainly meet the criteria!" "Er, thanks . . ."). "But then again, it could just be that life's hard at the moment." This is precisely what I was hoping he wouldn't say. Because what can I do about "life"? Go on strike? Run away? Hide, from my #CheckMyPrivilege hugely fortunate "life"?

I think of all the childless-not-by-choice men and women I know who would do anything to become parents. And here I am, moaning. Guilt crushes my chest, making it hard to breathe. How can this be?

A study conducted by sociologists George W. Brown, Tirril Harris and Maire Ni Bhrolchain[9] in 1975 found that one of the biggest risk factors for depression in women was caring for three children under the age of five. So if I'm not the only one feeling like this, perhaps it's something we ought to be able to talk about. Because if it becomes too hard to be a woman and a parent, might not this be a problem? Looking at the study on women with three under-fives, I can't help thinking that a few of them may not have been depressed: they might have just been bone-weary. Worn out. Exhausted from the mental and physical toll of the "second shift," the emotional labor of domesticity and caring, of lunchboxes, healthcare and vacuuming. Work without benefits, or a salary, or opportunity for promotion. Work that's been talked about by sociologists since the 1980s but that is still surprisingly hard to articulate today. We dangerously underestimate the demands and difficulty of combining work and parenting, then are somehow surprised when they are systematically sapped—trying to raise children in a nuclear family set-up, a set-up proven time and again to be wholly unfit for purpose and isolating [*steps down from soapbox*].

Dream doctor does not offer to babysit my life while I go off on a silent yoga retreat. Dream doctor does not offer to elope with me to a yurt somewhere. Dream doctor tells me, with by-now-watery eyes, that I have two options: therapy or drugs.

14

Going Pro

"Therapy" or "drugs"? That's it? Nothing involving a yurt? Or yoga? I'm hoping dream doctor will check his computer screen again and there might be a third option, like "a lovely long sleep." Or "a cyborg to help around the house." Or "a referral to the *Queer Eye* team." But no: it's therapy or drugs.

I'm in Denmark, currently ranked seventh out of twenty-eight OECD countries for antidepressant consumption.[1] Almost half a million Danes take antidepressants—in a population of 5.6 million[2]—so I'd be in good company here. And antidepressant use has tripled in the US since 1986.

More of us are on drugs than ever before. So are we really more depressed? Or are we swifter to diagnose? Hotter on prescribing drugs as the solution? Or all the aforementioned? Couldn't it just be that modern life is hard? That we're *all* experiencing burn-out?

"A lot of people point out—very validly—that what we're doing now is going to be considered ludicrous down the line, in twenty, thirty, forty, fifty years," says the neuroscientist Dean Burnett. "There are those who argue that depression and anxiety are actually healthy expressions of a mind that is commonly being bombarded by stress and novelty in the world we've created around us." The psychoanalyst Adam Phillips told the *Guardian* newspaper: "The reason that there are so many depressed people is that life is so depressing for many people."[3] Automation and the internet age were supposed to free up our lives for leisure. But for the most part, we now do more than our parents' generation. The digital world and all its freedoms mean that we're *always* on call. We shop online, then clash with customer service bots when things go wrong. We check out our own shopping at the grocery store. We plan and book our own holidays, checking ourselves in and printing our own baggage tags and boarding passes. Countless daily tasks that would formerly have been taken care of by someone else—the sales assistant, the milkman, the store checkout worker, the travel agent, ground crew—now come under our purview. The social drivers toward mental health problems are unmistakable. As Dr. Mark Williamson from Action for Happiness puts it: "In many cases, anxiety and depression are completely natural and understandable reactions to the reality of the situation you find yourself in." One friend has begun to find modern life so unbearable he's taken to cutting out screen time and wearing ear defenders and eye masks to dampen down stimuli from all around. I'd quite like to join him for a few hours a day, to counter the daily doses of stress that we usually take for granted. To reduce the need for daily doses of something else, instead.

I am offered the most commonly prescribed antidepressants, selective serotonin reuptake inhibitors or SSRIs. Neurotransmitters or chemical messengers communicate between the neurons in the brain and one of the most important of these is serotonin. Normally, when serotonin is released, it acts on receptors in the brain. If too much is squirted out, it's taken back up into the nerve endings to be recycled. But when we feel depressed, serotonin levels are lower. So SSRIs work by blocking this process. Instead of being taken back up, the serotonin gets to hang around for longer in contact with the receptor and so hopefully have more of an effect on our mood.

"For a while the central tenet of antidepressant theory was the monoamine hypothesis," Burnett tells me. "This is the idea that depression results from a lack or lower levels of certain types of neurotransmitter chemicals like serotonin." But in recent years, the monoamine hypothesis has fallen out of fashion. "It's insufficient," says Burnett, "because there is a lot more going on than just that. Maybe the depression is down to a lack of neurotransmitters, but why? Why are these neurotransmitters so deficient? What caused that? And before that, what caused that? The monoamine hypothesis doesn't really explain what's happening." The psychiatric geneticist Kenneth Kendler goes further: "The monoamine hypothesis is largely bullshit. The direct evidence for it is *very* modest."

SSRIs also raise serotonin levels immediately, but the beneficial effects on our mood take weeks to be felt. Why?

[*Science shrugs.*]

And if SSRIs can increase the amount of serotonin active in our brain and serotonin makes us feel good, then wouldn't we all feel fantastic if we took an SSRI?

"No," Burnett assures me.

In "nondepressed" people, SSRIs have been found to have no effect on mood. If we're not "depressed" and have no history of depression, then even lowering our levels of serotonin leaves our mood constant. It's only if we've got a history of depression or are currently suffering from depression that we're sensitive to SSRIs. The psychopharmacologist* Phil Cowen at Oxford University explains this seemingly cruel incongruity in terms of "scars"[4] on our brain: if we've suffered a pre-

---

* Scrabble winner . . .

vious depressive episode, the pathways in our brain are disrupted, making us more prone to depression in future.

So my brain is basically scarred for life with doom rivulets? Thanks, science . . .

"A lot of studies show that SSRIs are not the best antidepressants but they are the best tolerated, by the most people, with the fewest side effects," says Burnett. Accordingly, they're the most frequently prescribed. But there are others. "Some of the most potent ones that have the most significant effect on depression and depression symptoms also have the most severe side effects," he says. "These are so powerful they're normally reserved for things like treatment-resistant depression or people for whom depression is so debilitating that they'll say, 'Well, I'd rather have the side effects than the depression.'"

But the other thing about antidepressants is that they can be hard to come off. Withdrawal can be severe and long-lasting and some argue that the "serotonin deficiency" explanation for antidepressant effectiveness also normalizes the idea of lifelong use—since we're, understandably, terrified of a future that "lacks" serotonin.[5] As one woman on SSRIs put it in a 2019 randomized control trial: "I just need it. For me this isn't a psychological illness, it's physical. And my body isn't able to make enough serotonin, so I take the pill to supply it."[6]

Only half of participants in the 2019 study were willing to even attempt to discontinue antidepressant use. This mirrors my experience. Wary of painful withdrawal symptoms, I've taken antidepressants for longer than might have been necessary in the past, scared of a future "without serotonin."

So if the serotonin deficiency narrative is unhelpful, the monoamine hypothesis has been debunked and the ~~doom rivulets~~ "brain scars" theory is too disheartening to dwell on, is there another approach we can take? Sound the trumpets and roll out the red carpet because there is! [*Drum roll!*] It's *neuroplasticity*! This is the ability to form new physical connections between neurons. Like a big, brilliant spiderweb in our brain.

When we're depressed, Burnett explains, certain parts of the brain become, essentially, exhausted: "They lose the ability to adapt and change to whatever's going on in the environment," he says, "and we know that neuroplasticity in depression has been decreased." This means that our thinking becomes more fixed: "The brain isn't as able to change or adapt and depression endures," says Burnett. This may have been a protection mechanism historically (*Saber-toothed tiger just ate my whole family: it'll be me next. Perhaps now isn't the time to start on that new cave painting/grand tour . . .*") but it doesn't help us to get out of depression when we're in its grip. Antidepressants that increase neurotransmitters often increase neuroplasticity, too, so this may be why they work as they do, long after transmitter levels are raised.

But the interesting and new-to-me news is that while antidepressants can help form new physical connections between neurons, a drug-free approach can too. "Just as antidepressants can have an effect on neuroplasticity, talking therapies can potentially do the same thing by encouraging you to use different parts of your brain," says Burnett. "The system, the networks in place have been suppressed by whatever happened, but you can almost kick-start them with talking therapies like CBT [cognitive behavioral therapy] that gives a sort of scientific constructed framework. It isn't what works for everyone, but it does work quite well for a lot of people." He tells me about numerous papers citing testimonials of people saying they thought CBT would be "stupid and useless." "But then they try it and they end up really liking it."

There's also the psychoanalytic approach to uncover the root cause of a trauma. "Delving into the unconscious and trying to say, 'Why me? Why now? What is causing this?'" says Burnett, but he's more skeptical here. "There are plenty of instances where someone's mental health isn't based on a specific trauma: it doesn't have to have been a major event and so looking for one big cause isn't going to achieve much if there isn't one."

But at least with talking therapies there are no adverse side effects, I counter.

"Other than false memory syndrome?" *Touché*. "It happens," Burnett tells me. "You've got a qualified therapist telling you the trauma in your life that caused all this and that you must be suppressing it—and the brain is very creative. It can end up saying, 'Oh, well, maybe that was it . . .' Then you can create more problems."

And we shouldn't even think about self-help mantras or positive self-statements when we're feeling down. Those with low self-esteem actually feel worse after repeating positive statements about themselves, according to studies going back to 2009.[7] I ask the lead researcher on this, the psychologist Joanne Wood from the University of Waterloo in Ontario, why this isn't better known and why so many people still start their days with a "you're the best!" mantra and she tells me: "Some positive thinking proponents strongly objected to my findings. Second, people *want* so much to believe that positive self-statements work! They badly want to believe it, so-called experts tell them it works, and it makes intuitive sense that positive thinking should work, and most people don't realize the importance of scientific evidence." So we can't just tell ourselves to be happy when we're sad without making ourselves a whole lot sadder.

Depression itself is episodic. "You get bad and then you get better," is how psychiatric geneticist Kendler puts it. Studies from as far back as the 1890s show that if left untreated, depressive episodes normally last around six months. "But if your life is off track, I personally don't feel we should say to people, 'Just wait six months,'" says Nelson Freimer of UCLA's Center for Neurobehavioral Genetics. "I think it contributes to the stigmatization of depression. It somehow implies that you should just 'pull yourself together.'" And for many, six months of depression is unbearable. Freimer tells me how he trained in psychiatry thirty-five years ago and experienced early on in his career "the despair" that comes with depression: "I had one particular patient who had not been a lifelong depressive—he was a veteran of two wars and he was under my care in the hospital, where he and his family reasonably assumed that he would be safe. And he still managed to take his own life." A study from China also found that untreated depression returned with greater severity at a two-year follow-up point.[8]

There is hot debate on the drugs versus non-drugs approach to treating mental health issues. But this runs the risk of further "shaming" those who choose either option and misses the point. As Kendler says: "Depression is the hardest of all the disorders—it's as difficult as it gets." And if even the experts aren't sure why we get depressed and what to do about it, there should be no shame at all in those of us experiencing depression seeking help via whatever means are available. We're not "better" than other people because we *didn't* take antidepressants—or because we did. We don't become less anxious in a "superior" way than those who needed medication. It's awful for *everyone*. In this respect, at least, mental health problems are spectacularly indiscriminate.

This is why Burnett believes in the combined medical and psychological approach to depression: "antidepressants *and* talking therapies. I wouldn't deny that there's an overreliance on antidepressants, but if you've got the option of giving someone six months of CBT with a trained professional, or a packet of pills, then that's usually the option a resource-starved health service (whether tax funded, insurance based or paid for at point of use) will resort to. That's a practical problem rather than an ideological one." It may be that we're too quick to medicate. It is extremely unfair that talking therapies aren't accessible to many, many people. But therapy as a solution isn't quite a panacea either.

"Even if money were no object, we couldn't offer therapy to everyone," says Freimer. "In no place in the world do we have enough highly skilled therapists to provide it." In light of this, Freimer is looking for solutions elsewhere. He's the director of UCLA's Depression Grand Challenge, a ten-year research project involving 100,000 patients aiming to generate results that will help halve the current burden of depression by 2050, and to eliminate it altogether by the end of this century.

Bold, I tell him.

"I know," he tells me back. And yet, he sounds strangely confident: "We've got researchers from neuroscience, genetics, psychology, economics, engineering—all over. We're working to identify genes and environmental risk factors that play a role in triggering depression, *then* devise better therapies." He wants to dig deeper into how treatments like electroconvulsive therapy (ECT) and antidepressants work as well as using high-tech strategies to screen, monitor and treat people so depressive episodes don't escalate. So that normal "sad" doesn't tip into something more serious and depressive symptoms don't develop into clinical depression. He's looking at lifestyle interventions (more of this in Part III) as well as different options for treating depression beyond the current "antidepressants or sitting in a room with a therapist" set menu. Freimer points to the FDA's recent approval of esketamine—a nasal inhalant derived from the anesthetic and mood-boosting drug ketamine—as a landmark of sorts: "It's the first truly novel drug for depression in thirty years."

Once more I think: *Bold*.

"Plus there's now good evidence that psychological therapies delivered over the internet are effective," says Freimer, "and that's what we need: to employ treatments that are scalable and can treat those who haven't had access before." A noble goal, if ever I heard one.

For me, therapy is an option this time. And for that I'm thankful. There's a waiting list (of course) so I opt for antidepressants in the short term. And they work, kinda . . . I feel as though I'm at least doing something before I can start therapy: an undertaking I'm not exactly looking forward to. I've had "bad therapy" before as well as the good variety and it's an expensive, time-consuming, emotionally draining endeavor.

You might find a therapist you like first time by fluke—in which case, well done, you. Keep hold of them. But you might find someone with whom an hour a week in a stale room is as close to hell as you can imagine. You might feel as though you're coughing up $100+ an hour for painful silence punctuated by the odd, unsettlingly slow animatronic blink in a studio that smells of soup. Just for instance. Or to look at a woman who is 80 percent silk scarves and who holds her head just a little too far back, giving her a snooty air. Or one who gets your name wrong (Hi, "James"!).

I have never, I should say, reclined on a couch to do my talking. I won't wear mascara when a weep is on the horizon (even waterproof: it itches with prolonged crying). There's no need to bring tissues, I can divulge, since therapists buy these in bulk. But scheduling is key. On no account arrange any important interviews, meetings or parent-teacher appointments straight after a therapy session, and if it's a remotely sunny day, pack sunglasses to wear post-weep.

Few come to therapy as a first choice, writes the Jungian analyst James Hollis in *Finding Meaning in the Second Half of Life: How to Finally, Really Grow Up*.[9] There's usually, he's observed, a fair dose of denial first. We try to carry on as normal, doing the same things we've always done and magically hoping for a different outcome. Then we try distraction—work, sex, shopping and other addictive behaviors are all employed here. Before finally admitting and submitting to the issue that has brought us within a five-mile radius of a psychologist in the first place. Opening ourselves up to a stranger is a risky adventure. But it's one I resolve to take.

I find Marina Fogle a great advocate for talking therapy. She tells me how helpful she found it after her son, Willem, died. "Counseling revealed things I'd never have got to on my own," she says. "Like not being able to cry in front of the children. If you don't cry, it's like telling them they can't either. It's hugely important." Therapy, done right, is a judgment-free, confidential space to talk and be vulnerable. The comedian Robin Ince also found therapy useful, having experienced impostor syndrome and hypervigilance after a car accident when he was three that left his mother in a coma. "I spoke to so many neuroscientists and therapists who all said, 'If you've experienced a catastrophe at that age, it will affect you.' I just thought, Oh God, how tedious! Putting the way I felt down to my childhood just felt too easy. Too neat. And I didn't want to belittle others' pain—people who had gone through much more." This sounds familiar. But all the therapists Ince spoke to assumed that he was already in therapy and so eventually he took the hint and booked an appointment. "Of course I found the whole process very uncomfortable," he says. But he did it. "The point where I speak the least with another human being is in therapy," he tells me. "I don't worry about being entertaining. I'll be silent for fifteen minutes and then for five minutes I'll strike on what I'm going to share. My therapist is good: she knows if I'm putting on a performance. It won't be tolerated. I would worry about the repetition of ideas, but then I learned that in therapy, that's the *whole point*. Something that can seem minor to an outside observer can shape you," he says. "And we shouldn't be ashamed of expressing ideas of why we are how we are."

As the psychotherapist Julia Samuel writes in *Grief Works*, a therapist is the: "one person that you don't have to protect from how badly you feel. And with whom you can explore the same questions over and over."[10]

Convinced, I go to see a man. Let's call him "Casper." We go into his consulting room. *Be prepared to be cried on!* I want to yell. *I've got lots of tears and they're ready to go!*

On his wall, he has a printout of a *New Yorker* cartoon, by Robert Mankoff, showing a therapist saying to his client: "Look, making you happy is out of the question. But I can give you a compelling narrative for your misery."

I think perhaps we're going to get along. I will later Google this quotation and discover that I can, should I so desire, have this cartoon made into a shower curtain. A helpful daily reminder of life's messy reality if ever there was one.

"Okay, tell me everything," he says as I sit.

"Everything?"

"From the beginning."

So I do. We talk about my mom. We talk about my sister. I tell him that my first ever memory is from the morning the Very Sad Thing happened. How no one told me where my sister had gone, but that she wasn't there anymore and no one was happy. I tell him about my dad. "He's not a bad person. He's not a bad father, even. He just *wasn't one* for me." I talk about him leaving. At one point I laugh nervously.

"Why are you laughing? It's not funny."

"No," I say, because he's right.

He asks about my experiences of motherhood, after trying and failing to become one for so long. I tell him.

He tells me he thinks I like to be in control—the implication being "too much." Hotly defensive, I try to persuade him that I'm not a control freak: "I'm a control enthusiast." But he's not buying what I'm selling.

He guesses that anorexia was my addiction of choice, and not through any physical clues. "It's the one thing you want to get 'better' at. You were a perfectionist."

He asks about work and I tell him about writing alone in a room, for months at a time. He suspects that I find books and words easier than people: it's easier to write about a conversation than to be a part of it; it's easier to write about life than to live it. He tells me things I already know but that no one else appears to have noticed before: that I hide behind my work. That I use it to distance myself. To detach. He observes that I have a tendency to say "interesting" a lot and wonders whether I do this with friends and family.

"Of course . . .?"

He gently hints that I'm prone to rationalize, intellectualize and analyze, "instead of *feeling*." He thinks that I create emotional distance through busyness. That I use it as an anesthetic. He says I'm scared of intimacy.

There is an uncomfortably long silence. A dog barks somewhere in the distance.

"Maybe I just have healthy boundaries?" I counter weakly.

"Or maybe you're like someone in an old-fashioned bank with a panic button you can press to make a grid come down in case anyone gets too close?"

Double-busted.

I know rejection and how it feels. I know that people can fall out of love with you. And I don't like it. Casper gets this.

"Your dad leaving taught you that people cannot be trusted. Being abandoned by anyone affects your confidence. So now you're scared of trusting. Of getting too close."

I think back to previous relationships and remember feeling studied. Scrutinized, even, and finding this unnerving. That dizzy part when you're on the brink of falling in love has often felt more frightening than anything else. I can barely dissociate hearing the words "I love you" for the first time from a distinct sense of queasiness. I'd feel nausea—and worse, ingratitude that the words didn't make me "happier." Or that I couldn't say it back.

"But why?" I ask Casper after a couple of months. "Why has this come up now?"

"Because you're burned out. Because you grew up thinking you had to behave perfectly to be loved. Because now you have children who act out. Because they can behave terribly and *still* be loved. And that's hard for you." He explains, essentially, that my inner four-year-old is jealous as fuck. I feel ashamed and saddened. Envy of one's own children feels like the ultimate taboo. But it happens. More than we might like to think. The psychologist Philippa Perry writes in *The Book You Wish Your Parents Had Read*:[11] "It's not so unusual to feel jealous of our children." The antidote for parents, Perry believes, is to think about what life was like for us at the age we find our children most challenging.

That's it, I think. It's all the stuff I didn't deal with then. All the "sad" I pushed down and buried. Well, no more.

I feel a sense of relief at being seen. Heard. Acknowledged. Given permission to mourn and to finally be sad. Hopefully so that I can master the art before I mess my own children up (I'll just make other mistakes instead!). I won't pretend that the process of therapy is fun: it isn't. But it can help.

I sleep well that night. In the morning, my toddler daughter climbs into the bed and rests her head on my face. This is her preferred position at the moment. If she could, she would sleep like this, always. Preferably holding both my hands. I'm uncomfortable but I don't move. I don't want anything to change in this moment.

After a few more sessions, Casper tells me that he thinks I'm doing okay by myself for a while. Like Mary Poppins in crew-neck merino wool, he is gone. I'm scared of doing it all—"life"—without him. But I have to. And I can.

This is what a good therapist can do: get to know us, relatively quickly and—occasionally—better than we know ourselves. They ferret out trauma. They can hold a mirror up, then give us the tools to cope with what we see.

We all have to unlock our own box of difficult. If we're lucky enough to be able to get help with this, we should take it. And not feel guilty about accepting help, even if there are people worse off than us (there are always people worse off than us). If we think we might be depressed, we should see a doctor (even if they're dreamy). And if we think we might need ongoing help but we can't access therapy, we can still get support. We still need to talk about our "sad." If we can't go pro, then we need a Jill—a buddy. Someone willing to turn the radio off if we've been tuned in to Shit FM for more than a couple of days. Let me explain . . .

15

The Buddy System

IT'S 9 A.M. ON A Tuesday and the phone exchange goes something like this:

Me: *Shit FM playing for two days now.*

Jill: *Coffee?*

Me: *In twenty minutes?*

Jill: *Thumbs-up emoji, flexed muscle, cup of coffee.*

I put my phone in my bag, safe in the knowledge that in twenty minutes Shit FM will get scrambled.

Shit FM is the term friends and I now use to refer to our inner monologue: that unhelpful radio station in our heads that tells us everything—and everyone—is garbage. Including ourselves. Shit FM has a wide and varied playlist, from melancholia and nagging dissatisfactions (let's call these "Coldplay" problems) all the way up to convulsing anxiety issues that make you want to tear your own hair out in frustration (thrash metal). We all tune in to Shit FM from time to time. But if the dial gets stuck—or if the tuner goes rogue and takes us back to Shit FM more than usual—then it's time to enlist help from a mood-engineer. A friend.

After the Shit FM call-out, there follows a caffeinated powwow on the woes of the world, or, rather, the woes of us. When she talks, I listen. When I talk, she listens. Neither of us tries to "fix" the other. We just work out how we're feeling by talking about it, out loud, to trusted ears. After which we both feel lighter. The "sad thing" hasn't gone away, but it has been aired. Like an old coat. It's a system I've found increasingly helpful after Mary Poppins/Casper disappeared from my diary as a regular feature. I knew that this time, I really (really) couldn't handle any future "sad" on my own. So I roped in some buddies in what I am now referring to as the *How to Be Sad* Buddy System.

The Buddy System originated in the US Army to keep troops safe when executing tricky military maneuvers, and the first recorded use of the term was in the 1920s, according to Merriam Webster.[1] In scuba diving, members of a group work together and cooperate with each other in "buddy diving," so that they can help or rescue each other in case of an emergency. Quite apart from this, the fun footwear and cool salt-tousled hair, scuba divers have another ace card we can all aspire to. The most widely known rule of the sport is "Never dive alone"—a motto that could equally be applied to life, replacing the "d" with an "l" to make it: "Never live alone." The "perfect buddy" in scuba diving is a long-term friend or acquaintance whom we could trust with our life—which isn't a bad idea for our land-based endeavors either.

The Buddy System is also used in that other high-octane arena of life and death: the Scouts. Cub Scouts, Girl Scouts and Boy Scouts worldwide use the Buddy System to help children learn to work together, start fires, sell cookies or avoid getting lost in the woods (depending on your troop). Because, as Barbra Streisand knew only too well, people need people.

"For years mental health professionals taught people that they could be psychologically healthy without social support, that 'unless you love yourself, no one else will love you,'" writes the American psychiatrist Bruce Perry in *The Boy Who Was Raised as a Dog: And Other Stories from a Child Psychiatrist's Notebook*.[2] But the truth is, Perry writes: "You cannot love yourself unless you have been loved and are loved. The capacity to love cannot be built in isolation."

Plato and Aristotle both recognized that friendship was fundamental to being human. It turns out I am not a rock: I'm not an island. And neither are you. We need to talk about things, especially when we're sad. Even the psychotherapist Julia Samuel agrees that talking about sadness "does not have to be with a therapist." What's most important, she says, is "talking to someone who doesn't interrupt." This helps us to develop a narrative around our situation and, as Samuel explains: "in annunciating the *words*, the *feelings* emerge."

If there's something on our mind, we should share. Preferably not on social media in a long Instagram post or in a cryptic Facebook status update, but to an actual human being. A buddy. A friend who we can check in with regularly. Although for many of us, this is easier said than done.

Talking in real time can feel scarier than commenting on social media or sending messages via WhatsApp, and "social anxiety" is a growing problem. An estimated 15 million Americans have social anxiety disorder, according to the Anxiety and Depression Association of America.[3] Telephone phobia (or "telephobia") is the reluctance or fear of making or taking phone calls and although it's been around for almost as long as there have been phones (the poet Robert Graves wrote about a fear of using the telephone in 1929),[4] cases are officially on the rise.

Anxiety over the prospect of talking has increased as we've moved away from verbal communication and toward texting, emailing and social media comments. When we compose a digital missive, we have time to think and edit and perfect. In conversation we're talking in real time. I'm often plagued by the idea that if I make a "mistake," it's "out there" and there's nothing I can do to take it back (see "recovering perfectionist").

When I started writing this book in 2019, many experts feared that greater access to technology could exacerbate social anxiety still further, particularly as smartphones, tablets and computers become omnipresent. Then COVID-19 hit, a lockdown ensued and we all had to learn to connect digitally. And we did. Zoom and Microsoft Teams all went from being something only the tech savvy or young were into, to being something even grandparents embraced. We all connected, digitally, for so long that it was a little strange and I felt shy to meet up again in person. But we managed this, too. Because like a muscle, we can train ourselves to be more resilient in social situations and connect via what means we have available.

Most therapists recommend exposure therapy—where we literally put ourselves through the thing we're scared of in small doses, building up slowly until it feels *fine*. So if we're scared of talking on the phone, we should try a thirty-second phone call first. Then build up to a minute. Then two. Before gradually increasing our minute count to a chat-appropriate call length that feels right for us. We should also take heart from the fact that we overestimate how much we're ever [*actually*] messing things up in social situations. According to research led by Professor Thomas Gilovich at Cornell University, we significantly overestimate how noticeable our embarrassing behaviors are to others.[5] We also underestimate how much our conversation partners like us and enjoy our company—an illusion called "the liking gap."[6] After we have conversations, we are liked more than we know. And if social media ever gives us the impression that everyone else is out there living their best lives, remember this: social media gives us an "impression." It's not "fact."

At the time of writing, no one quite knows how the isolation necessitated by COVID-19 will impact us, long-term, but some predict that the next pandemic may be loneliness. A National Academies of Sciences report[7] found a consistent relationship between social isolation, depression and anxiety, and researchers from Washington University in St. Louis suggest that loneliness may be a forerunner to mental health problems.[8] A lack of strong social connections has been linked to poor diet, heavy drinking[9] and high blood pressure—and could be even as damaging to our health as smoking.[10] A meme goes around the internet at the start of lockdown, illustrating that during times of isolation we tend to turn to food, alcohol or exercise as a crutch (pick your poison: "chunk," "drunk" or "hunk"). Having fallen for all three in the past (hard), I am wary. As the philosopher and former addict Professor Peg O'Connor puts it: "Connection is important in the context of addiction—of living in recovery or moderation management." She says of the pandemic: "With the isolation, the fear, the panic, the anxiety, the melancholia—all of that—I think we'll see spikes of people beginning to go down that road of addiction. And I bet we'll see spikes in relapses, too."

I find the only thing that helps me during this time is the daily WhatsApp chats with friends and regular FaceTime calls. It's not ideal. There's none of the oxytocin of a good hug or the joy of human contact in person. But it's what we have and I always feel better afterward.

We can't turn back the clock: smartphones are part of our lives now (an extension of our hands, in some cases). And as well as creating hang-ups about actual, in-real-life conversations, technology *can* be used as a tool for facilitating more meaningful interactions, too. Like the rainy Tuesday Shit FM text. Like sending a virtual "Bat light" into the air to let the people who matter know we need them. Like pressing a button to let our closest friends know when we're in trouble.

This was the thinking behind the notOK App—a free mental health

SOS app developed by teenage siblings Hannah and Charlie Lucas in Georgia. After developing POTS, a condition that causes her to faint, then fifteen-year-old Hannah Lucas became terrified of being alone. She worried about what would happen to her if she fainted and no one was around and her fears grew into anxiety, depression and self-harm. After a suicide attempt, she told her mother: "I wish there was a button you could press and someone would know when you're not okay."[11] Her then eleven-year-old brother had been learning programming since he was seven (#overachievers) and so started working on a wireframe for an app. Together, they took coding classes and wrote a fifteen-page pitch to their parents for funding to hire developers ("We even allocated ten minutes at the end 'for questions'"). Their mom helped them find developers willing to work out a payment plan and six months later they had an app. The notOK App allows users to alert five trusted contacts—five buddies—at the touch of a button when they need some extra support.

Why five? I ask.

"We figured that five was a big enough number that you could still get the help you needed if someone was busy," Charlie tells me over Skype, "but not too many that you got social anxiety about not having enough friends!" All hail Gen Z.

The notOK App works by sending a text to five pre-agreed friends with the message, "Hey, I'm not OK. Please call me, text me, or come find me," along with a link to the user's GPS location. "And it works," says Hannah. "I got help when I needed it." She felt instantly safer and supported. "The app gave me a sense of comfort," says Hannah, who thought others might benefit from it, too.

The app launched on Google Play and the App Store in 2018 and has now been embraced by those suffering from eating disorders, anxiety and addictions, too, with 87,000 users. Now, the siblings are working on notOK App 2.0 and have been awarded grants to expand globally (visit notokapp.com for more). "The reaction we've had so far has been really positive," says Hannah. "Some people don't have the words to tell somebody they're not OK but they still need a way to reach out for support. So we wanted to make that connection a little easier." Because we all need buddies and peer support has been shown to improve quality of life and our health, according to data collected by Mental Health America.[12] The Harvard Study of Adult Development, started in 1939 and continuing today, states categorically that, "Good relationships keep us happier and healthier. Period." And Charlie's hunch that five might just be the magic number was spot on.

The evolutionary psychologist Professor Robin Dunbar from Oxford University has conducted years of research into social bonding and found that we need five close friends, fifteen good ones and up to a hundred and fifty in our outer circle. "Each of these layers corresponds to a particular level of emotional closeness and to a particular level of frequency of contact: the limits on the layers are at least once a week, once a month, about once a year," Dunbar tells me. "This may be because creating bonds of a given emotional intensity requires the investment of quite a specific amount of time. Drop below that, and the person slips very quickly—within months—into the next layer below in terms of emotional closeness." One of the biggest barriers to meaningful bonds is distance, with thirty minutes being the longest time we'll spend traveling to meet up with a buddy, according to sociologists Barry Wellman and Scot Wortley.[13] "It doesn't matter whether it's on foot, by bicycle or car, thirty minutes' travel time is the tipping point," says Dunbar. This research was conducted pre-pandemic, so while five close friends who we see once a week and only live thirty minutes away is the goal, the general advice during any sort of global/personal crisis is: "Do your best."

Five friends we see weekly sounds relatively straightforward and accessible but can feel like a big ask in our busy, modern lives. And many of us don't have this. A third of Americans over the age of forty-five say that they're "chronically lonely" according to a 2018 study.[14] There's even a UCLA Loneliness Scale—a twenty-item assessment designed to measure subjective feelings of loneliness and social isolation.[15] Experts have discovered that key life events—even ostensibly happy ones—seem to spur a spike in loneliness. Thirty-two percent of new parents say that they "always" or "often" feel lonely, and I remember the isolation of early motherhood vividly. The end of a relationship often marks another, less surprising, spike in loneliness, with 33 percent of those recently divorced or separated saying that they "always" or "often" feel lonely and more than half of those recently bereaved reporting that they wouldn't know where to turn for help. Many of us are lonely and we don't know what to do about it. Because forming new bonds to replace ones we may have lost or that are lacking isn't easy.

If we relocate for some reason as adults, it can be tough to make new, local friends. My mother moved in her sixties and had to start over from scratch. The energy and effort she had to expend in order to kindle a new social circle was exhausting. My mother is a sociable creature but still, she found it hard.

Traditionally, many of us who are forced to make new friends as an adult may have done so at work. But the gig economy, working from home, hot-desking, and even open-plan offices mean that it's more challenging to make meaningful connections—to make *friends*—in the workplace than ever before. Even some new residential developments, built to maximize individual privacy, mean that we don't walk past our neighbors on a daily basis anymore. When we lose this, we miss out on much more than just over-the-garden-fence/communal-wall gossip.

I have good friends I haven't seen enough of lately, but as Casper and Barbra Streisand have all reminded me: I need people. I need my buddies. And, I ~~think~~ ~~hope~~ *know*, my buddies need me, too. Because many of them are beginning to behave a little strangely. Just before COVID-19 struck, my peers and I hit a milestone birthday . . . and various strands of our lives begin to unravel.

This Is 40, the 2012 movie, gave me the distinct impression that although the start of middle age *may* encompass money worries, difficult teenagers and surprise pregnancies, we would all have thick, lustrous hair (like Paul Rudd and Leslie Mann) and be gadding about in Californian sunshine.

This Is 40, the reality, involves money worries and surprise pregnancies (tick) but no one I know has hair that good and, so far, it rains a lot. *This Is 40*, the reality, also includes chronic illness, bereavement, a global pandemic and mental health problems as, one by one, people around me slide into a spiral of clammy despair. We need friends at every age. But I'd hazard that we need them even more once the formal festivals of friendship have fizzled out. For years, we see friends daily—at school or college. After that, there are often weddings, bachelor or bachelorette parties. By the time we hit thirty we may have a solid social network and relatively few responsibilities, so we can hang out. Gather. Connect. Have time for four-hour-long, in-depth conversations. Or nights that turn into lost weekends. Then the nuptials thin out and other commitments stack up.

By forty, we may have less time to see our friends but a greater need for them than ever before. Most of my friends work too much, have responsibilities outside work, and then somehow manage to pile on extra pressure in their so-called leisure time. We're then surprised when things start to go a little haywire. A few become evangelical about HIIT or weight training. They lift barbells and say, "Hnnnnnnnhgh!" at their own reflections. A lot buy bikes. Some get tattoos. At the last count, we had five tattoo parlors in our town along with an alarmingly ill-thought-through "pop up" ink bar on the street. Some forgo ink for illicit affairs. One friend leaves her partner for a new one who's very . . . *lookatable*. This is fortunate, since they don't speak the same language. Another has an affair with a man who could not look more like a murderer (think early eighties serial killer) if he tried. And that's just the women.

A male friend in his forties takes up competitive-level drinking with such exuberance that his wife catches him taking a dump in the front garden in the early hours one morning ("*Why?* Why would you do that? By the *begonias*!"). Another asks his wife for a divorce after two bottles of Merlot. She tells him: "You can't have a divorce, what you *can* have is the guest room." In the morning, he can't remember a thing. Another becomes a recluse. Friends try to get him to come out of his room, but all he wants to do is lie in bed and eat pudding.

What is going *on*?

"It's the classic 'midlife slump,'" the journalist Matt Rudd tells me. He's five years ahead of me on this particular adventure (he's forty-five) and tackles the topic of midlife in his *Sunday Times*'s Man Trouble column. "On the surface, middle-aged men are quite low down on the pile of people we feel sorry for," he says. "You know, we're the privileged ones. Everything should be fine." He's right. It should. These are, it's necessary to point out once again, first-world problems and if 2020 taught us anything it was that white middle-aged men have been inhabiting a space of privilege since time immemorial and don't have much to moan about. And yet . . . Since Rudd started the column in 2019, he's spoken to hundreds of "apparently successful" middle-aged men, living in nice houses with supportive wives and 2.4 children, "and they're all miserable!"

In our twenties we have "potential." In our forties we probably should have met it. If we haven't, we feel bad. If we have and we're still not satisfied, we feel bad. We're on the hedonic treadmill—the established tendency of humans to quickly return to a relatively stable level of happiness despite major positive or negative events or life changes. Or it's the u-shaped curve whereby midlife means an inescapable dip in happiness, according to experts. But Rudd has another theory: "To 'fix' this we'd all have to 'go back in a time machine and not give a shit about school.'"

He explains: "The way school is, from a very early age it's all about winning. It's all about reward for being well behaved. From a really young age, they are in a classroom having to sit still. They attach good behavior to success." From my research, a similar approach is taken in much of the world. "There's no space to think, Will this thing really make me happy? I mean none of this is *Angela's Ashes*–type stuff, but the pressure . . ." says Rudd. I nod. Parents want us to succeed; school wants us to succeed; and few of us are urged to challenge this. On leaving formal education, Rudd says, "you feel like you've done everything that was asked of you, but you're still not happy." So you get a job ("still not happy"), enter into a relationship (ditto), and perhaps, even, start a family.

"And then childbirth is another insane part of how men are expected to just continue on!" says Rudd. Expectations around fatherhood have changed dramatically in the past forty years (thank goodness) and Rudd has three sons with whom he's tried to be "hands-on" from the start. "But I could only take a week's paternity leave for each of them," he tells me.[16] "When Freddie [his eldest] was born it was a very difficult birth—really touch-and-go for the whole fifty-four hours of labor. It started as a home birth and ended up in emergency Caesarean. Three days later, I'm back in the office. And then I've got to continue with my patriarchal role as provider—which men still feel—but also to get home and switch into 'dad' mode. You're just plowing on," says Rudd, "so it's no surprise to me that when you get to forty you suddenly think, Was all of this the right thing to be doing? And so many of them think, Well, *was* it worth it? No."

I hear him, although confess that my female friends are also going a little nuts.

"But at least they talk about it," is his response.

This is fair.

"Men don't talk about emotions," he insists, "still."

They haven't yet adopted the Buddy System.

"I think every time I've got into a really intense conversation with a middle-aged man, they have always at some point got quite nervous, because they don't want to think about their happiness. They don't even want to discuss it. Because they then become worried that the wheels will fall off. They worry that they won't be able to get out of bed in the mornings," says Rudd.

I wonder whether this is down to how many men are raised and how their parents approached conversations about painful or uncomfortable topics. I ask Rudd if he ever talked to his dad about emotions. He shakes his head. Hard.

"I have *never* had an in-depth conversation about feelings with my father, it's very traditional." Rudd assures me that the clichés of locker-room talk that I grew up watching on TV are still depressingly familiar. "Whenever you're discussing something serious, someone will always defuse the situation rather than work through it," he says. "So, one friend is explaining to another friend that he's got cancer and within three sentences, he's doing a testicle joke. That happens a lot."

Ah, humor as a defense mechanism! Hello again.

Rudd is guilty of this, too. "Every time I've been sent off on some course I've been told that I 'must not use humor to get out of difficult situations.'" He passes a finger around the rim of his collar, as though overheating at the memory of such discomfort. "But I ignored all of that because, well, funny *is* quite an easy way of avoiding revealing too much of yourself," he says. "But men need to be able to talk, and I'm really passionate about that."

So is Richard Clothier. Remember Richard? The spokesman for male infertility who desperately wanted a baby with his wife? Well, they've had some good news. The second cycle of IVF worked. Which is, obviously, wonderful. Although I asked him if he has what I have: the guilt. That we've somehow "defected" to the #asaparent side.

"Yeah, 'survivors guilt,' definitely," he says. "Of course, I wouldn't swap being a father for anything but at least I'm mindful of the struggle of male infertility now. I only talk about how happy and excited I am to be a dad now with very few people. And I think what's more important is that we get better about talking about our feelings—especially as men, especially with our friends. Otherwise, what are they for? Sometimes men are our own worst enemies. So now I open up about things—because I know firsthand that it really does help."

Londoners Jack Baxter, twenty-eight, and Ben May, thirty-three, have also become champions of the Buddy System. "In an ideal world," Baxter tells me, "Ben and I would only be friends because he cuts my hair and we both love [soccer]. But actually we met because we both lost our best friends and joined a club no one wants to be a member of—the Dead Dad Club." Baxter went in for a haircut at May's barbershop in 2015 when his father had just died of skin cancer and May's dad was dying of a brain tumor. "We talked about [soccer] first, of course," says Baxter, "but we soon started talking about our dads. And it turned out we'd both been brought up by these really old-school masculine men who liked sports and drank beer."

Baxter's dad was a bodybuilder. "He was 250 pounds and only cried a handful of times his whole life. He was tough. He actually said that his weight training helped him get through the cancer treatment because he was prepared for the pain. But I wasn't." He remembers feeling fury at the injustice of it all: "We never talked about relationships or emotions growing up, so when my dad got sick, I didn't know what to do with these feelings. I just had this rage." He describes himself as "an angry young man, looking for a fight. . . . I wanted someone to blame." Baxter works in TV, "a world full of six-foot-tall men wearing suits who don't 'do' sad. It's a world full of ego—but there's no room for ego with grief. And there didn't seem to be anywhere I could talk honestly about my emotions." Baxter had been influenced by the only model of "masculinity" he'd ever experienced, the "toxic" kind, "you know, the idea of masculinity that said 'real men don't talk about feeling sad.'"

May can relate: "My family didn't talk about emotions growing up, either—we'd mainly just joke around with each other." Interestingly, May spent eighteen months in anger management before his dad died. "And that helped," he thinks now, "because when Dad died, I felt this overwhelming grief—but at least I *knew* it was grief. I wasn't angry at people in the street for no reason. When it happened, I understood that I was actually really sad."

As a barber, May now gets customers coming to him specifically because they know that they can talk to him about sadness. "Especially men," he says, "who are often still wary of expressing themselves." May cuts around fifty people's hair a week and few pass through his door without unburdening themselves in some way. "We might start off talking about [soccer]," says May, "but you attract people based on what you put out there. And now people are familiar with my story—through social media—and realize it's something that I will talk about. They know that I will probably mention my dead dad at some point."

Does this ever make customers uncomfortable? I ask.

"I don't care if it does!" he says. "That's tough, we're talking about it. It's important. Because feelings matter and they're nothing to be ashamed of."

Baxter agrees: "We want to be the next generation of men who are happy to talk about our emotions and cry in front of people."

And the barbershop seems a fitting place to start. A study from the Samaritans found that hairdressers and barbers spend an average of 2,000 hours a year listening to their clients share their troubles.[17] The idea of barber as confidant and the salon as a "safe space" for men to talk is nothing new, but it's now being formally acknowledged as something fundamentally worthwhile for men's mental health.

Lorenzo Lewis founded the Confess Project (theconfessproject.com), a countrywide nonprofit organization where barbers can connect with men of color and raise awareness about mental health. Through a twelve-month curriculum, barbers are trained in active listening and how to use positive language to combat the stigma around mental health. The mental health campaigner and barber Tom Chapman is developing a clinically backed training program—BarberTalk—for hair stylists to support their clients and communities. Chapman founded the Lions Barber Collective to help train over 13,000 barbers, potentially reaching 2.2 million people every week with messages of hope, after a close friend died from suicide. One person dies every forty seconds from suicide globally, according to World Health Organization data[18] and 78 percent of all suicides in the US are male.[19] In the African American community alone, suicide is the third leading cause of death.[20] Men are three and a half times more likely to die by suicide than women in the US[21] and for males under forty-five years old, suicide is the second most common reason for death (the first is accidents).[22] Yet men aren't historically great at asking for help when they're struggling. Research has long shown that men tend to visit the doctor less often than women,[23] and a 2019 Cleveland Clinic study found that most men would rather do almost anything—including cleaning the bathroom or mowing the lawn—than go to the doctor.[24] When they do go, they're less likely to be honest about any problems.[25] A study by psychologists Diana Sanchez and Mary Himmelstein published in *Preventive Medicine* found that men with traditional ideas about masculinity—who believed that men should be "tough," self-reliant and restrained in terms of emotion—were more likely to ignore medical problems, or put off dealing with them, than women or than men with less traditional beliefs. They were more likely to choose a male doctor (because: "Women . . . amiright?") but were less likely to be open with that doctor about their symptoms (because: macho).[26] Many report feelings of shame around admitting mental health struggles. But talking and listening saves lives—an emotional win-win.

I am far from being an apologist for male privilege. But it seems to me that we're a long way off from achieving equality if we're not giving men equality of emotions, too. (I'd also be willing to share the gender pay gap, menstruation, childbirth and the menopause in case anyone's interested . . .)

I talk to T about this and he agrees in part: that there are some nights he lies awake worrying how we'd pay the mortgage if he lost his job (you'd be hard-pressed to feed and clothe a family of five on a writer's income in Denmark).

T is pretty good at opening up (as my Canadian mother-in-law assured me when we first met, "This one wears his heart on his sleeve!"). But still, he admits, "it's a dance: a navigation" that's needed with male friends as to whether or not you can open up without fear of judgment. Or worse: mockery. "There is a leap of faith," T says, in trusting someone. "It's like, you throw all the jigsaw pieces into the air and watch them scatter."

But you still do it? I ask. You still put yourself out there?

He nods.

Why?

"If you don't ask, you don't get." He shrugs.

Quite.

16

Support Network Needed

W<small>E COULD ALL</small> do with asking for help more and being honest about our vulnerabilities—opening up and actively seeking out "buddies" who can be our support, as we can be for them. And to truly cope with being sad, well, we need a network. We have to talk about our problems and how we're feeling with people we can trust.

Many of us struggle when discussing mental health issues, stress or weakness. Our education, training or vocation can also influence our proclivity to open up, with one group in particular deserving of special mention. A group for whom the stakes are even higher.

Add "expectations of infallibility" to "stoicism on steroids" and we get the perfect storm that is . . . the medical profession. Adam Kay's diary about his life as a junior doctor in *This Is Going to Hurt*[1] is hilarious and heartbreaking in equal measure as he documents just how hard life is for junior doctors, coping with life-and-death situations with very little support. I'm interested in how doctors do "sad" so I speak to Kay. He confirms that the "support" part is sorely lacking. It's hard on personal relationships, too. Kay is no longer with the partner he talks about in *This Is Going to Hurt*. "Relationships break down for very complicated reasons," says Kay, "but the job is certainly a factor. I'd be interested to hear about the divorce rate among doctors."[2]

T's ex-girlfriend was a doctor. A wonderful woman by all accounts, but one who could apparently run low on empathy after a hard shift in surgery. Understandably. Kay gets this: "Doctors are walking a tightrope: you don't want a doctor to be an emotionless psychopath, but then you also don't want one who'll burst into tears when breaking bad news. So most doctors are on the continuum of being too hard as opposed to being too soft. Compassion fatigue is definitely a defense mechanism." That, and dark humor. Kay's book has abundant tales of things stuck in orifices where they shouldn't be (including, memorably, a TV remote) and there's an age-old marriage of comedy and medicine. "Telling funny anecdotes is a universal," says Kay, "doctors need to find a way to cope and funny stories help with that." Humor is a healthier coping mechanism than many of the alternatives adopted regularly within the profession. He tells me how many medics drink a little too much and tend toward recreational drugs in their spare time: "Everyone is desperate to find a way to cope." For Kay, it was writing in his journal.

And did that work?

"It was absolutely not *enough* of a coping strategy," he says, "but it papered over the cracks along the way. Until it didn't. And I couldn't cope."

The hardest part of Kay's book comes in an entry from December 5, 2010.

Kay began to perform a Caesarean on a patient with undiagnosed placenta previa—when the placenta is in the way of the cervix and has to be delivered first. The baby died. The mother lost twenty-five pints of blood and needed a hysterectomy. None of this was Kay's fault ("All my peers would have done exactly the same thing and had exactly the same outcome," he writes) but he couldn't come to terms with what happened and he felt bereaved. He couldn't laugh, or smile, for six months.

Kay now reflects that he should have had counseling: "But there's a mutual code of silence that keeps help from those who need it most." Kay didn't talk about what happened with anyone. Friends and family thought he was having a breakdown. Which he was, in a way. He tells me that he had flashbacks to what happened for years after: "It became a feature of my sleep cycle—really, until I started talking about those events." And he didn't start talking about them until the book came out. Seven years later.

"Initially, talking about it was a difficult thing to do," he says, before correcting himself. "It *remains* a difficult thing to do. It's acutely painful. But then the flashbacks stopped." Friends and family finally found out the real reason he left medicine.

How did they react?

"They were angry," he says, "or rather, they were *disappointed* that I felt they weren't approachable enough for me to have talked to them about it. But it was absolutely on me: I chose to bottle it up." Then he adds: "I say it was on me; it was also the way I had been trained as a medic. To toughen up." Because doctors aren't supposed to show emotion. Feel emotion, even, in much of the world (my misty-eyed Danish doctor aside).

Kay says that he still gets the odd person messaging him or coming up to him after a book event to tell him to "toughen up." "I've had people telling me I'm histrionic, too," he says.

I wonder out loud whether these people have been male or female.

"Exclusively male," Kay tells me. *Uh-huh.* He also has male doctors getting in touch to encourage him to "grow a pair." "They honestly believe that's the way a doctor should be," he says. "I'll talk about ways to look after junior doctors and they'll say, 'Yeah, I hear what you say, but doctors really just need to absorb the shock.' But if you're absorbing shock after shock, it will eventually damage you, that's the thing. It's death by a thousand cuts."

To deal with the "shock" and cope with distress and sadness now, Kay tells me, he's had to "unlearn" some of the things medicine taught him: "I've had to learn instead that it's good to talk about stuff that is worrying you and upsetting you. I have to remind myself that actually it's not weak or weird to say to your other half, 'I've had a bad day.'"

The way doctors are supported also has to change, says Kay. "You need an ongoing network, a team looking out for you, so that you know where to go when things aren't going so well. And then in a more acute setting, when the very bad things happen, you need a safety net to pick you up." In laypeople's terms: you need a buddy.

Kay's book has become a phenomenon—selling 2 million copies. But one of the most extraordinary things is how Kay's story has resonated internationally. It's now translated into thirty-seven languages.

"People all over the world have said it could've been set in their hospitals," says Kay. "Ultimately it's about the experience of being a doctor. Which is the same wherever you are."

I talk to my mother about my interview with Kay. She's an avid fan of anything medical and has watched so many episodes of *Grey's Anatomy* that she fully believes she could deliver triplets single-handedly, should the occasion arise. Naturally, my mother has read Kay's book. She's also taking a special interest in my research for this book. She calls it "our story"—and she's right: it is. So she's invested in what's going in it.

It's one of her last visits before the coronavirus lockdown, although we don't know it yet. But we're talking, properly, which is new for us. We chat as we peel vegetables for dinner one afternoon—her on potatoes, me on carrots. We're perched on toddler high chairs at the kitchen table since all counter space is taken up (burned pans, a lone oven mitt, half-eaten apples, several yo-yos and a wrench, mostly) and the normal, grown-up chairs are loaded with books, soft toys or encrusted oatmeal from a bellicose breakfast.

I tell my mother about Matt Rudd's theory, that school sets many of us up poorly for the world. I tell her about how having a "buddy" might help all of us handle the tough times better. About Kay's experiences and my theory on doctors, and boarders, and denying our emotions as a coping mechanism—a hangover, possibly, from wartime.

My mother nods and says: "That makes sense." She's digging a particularly stubborn eye out of a potato when she tells me: "Your grandfather was in a German prisoner-of-war camp in Italy during the Second World War and saw that the only other men who got through it were boarding school boys." I'd known he'd been in a prisoner-of-war camp, but it was never spoken of [*plus ça change*]. I hadn't known the part about the boarding school survivors. "Apparently they were better able to tolerate the conditions," my mom adds. I'm reminded of the historian Thomas Dixon writing about how "resilience" was prized and solidified during the war. My mother flicks out the eye of a potato expertly with her peeler and it lands in the middle of the table. "Your grandfather swore if he ever made it out alive and had sons, that they'd be boarders," she adds casually.

"Wow. Right. But, wait, he *did* have a son?"

"Yes."

"My dad . . ."

"Yes."

"So my dad went to boarding school?"

"Yes."

I am momentarily dumbfounded.

"Didn't you know?" She looks up at me.

How would I know? No one in my family talked about anything! Ever! How would I know?

She sets down her peeler, wipes her hands on an apron, adjusts her spectacles and uses her index finger to Google the school he attended on her phone before showing me its online prospectus.

It looks isolated. The pictures show no trees but I somehow still get the idea that it is windswept. It is also grand. I read that the school has an illustrious history dating back to the 1500s. My family does not have an illustrious history (no offense, family). My grandfather must have scrimped and saved to send my dad there.

"And, how old was he when he started boarding?" I ask.

"Eight."

EIGHT?! To be separated from family—from everything familiar—at such a young age is hard to imagine. I think of my grandfather anticipating a future where sending his eight-year-old son away was deemed necessary for his survival. I think of a generation that came of age during war. A generation who believed that what wasn't spoken about wouldn't hurt you. And so sadness wasn't spoken of. My grandfather didn't talk about it. And his children didn't, either. And now his children's children are starting to think that it *might* just be worthwhile, this "*talking*" business.

I wonder what my mom thinks, now, of all that went unspoken.

"Now? I'm not sure that made the pain any lesser. Now, I think it might have made it worse."

Talking about sadness is often so off-limits that many of us struggle to find the words to try, even when we want to. And if we somehow manage to find the words, it can still be hard to find someone who wants to listen. This was my mom's experience in the early eighties when she found herself alone, grieving, and with no one to talk to.

"People didn't want to hear about sad things. About death. It made them uncomfortable. So it was lonely."

She tells me about a time the week after my sister died when, after several days of house arrest, she summoned all her strength to put on her blue polka dot overalls (see "1980s") and take me to a mother and toddler group at the church hall. As we walked in, silence descended. The other mothers parted before her, like she was Moses in a pair of blue spotted overalls. No one looked at her. They cast their gaze down, hurriedly, as though direct eye contact could contaminate them in some way. As though they would be "infected" with sadness. A few of the children were bolder and let me play with them, apparently, but my mother stood alone. Shunned by the women who just last week were lining up to coo over her now-departed baby. "We left and walked home again soon after that," she tells me. My mother picks up her peeler and starts on another potato, before adding: "I think that's why I became so fixated on the Whirlpool man . . ."

Wait, what? Whirlpool as in the washing machine brand? I try to follow my mother's breadcrumb trail: *boarding school, my dad, emotions . . . Whirlpool?*

I am none the wiser. So I ask.

"I never mentioned the Whirlpool man?" She sounds surprised.

I tell her, with some certainty, that she has *never* mentioned the Whirlpool man.

"Oh, well then . . ." She begins to explain. My mother tells me how she'd had trouble with the washing machine during her pregnancy with my sister and that the same man from Whirlpool had come to look at it a few times ("In those days you got *real* customer service"). Then, when my sister was born, the machine broke down again ("And newborns produce a lot of laundry!") so Whirlpool man returned.

"I remember he held Sophie and played with her while I made him a cup of tea." Then he mended the machine. Then my sister died. My dad went back to work and my mom was on her own, the other moms at the toddler group having cold-shouldered her, terrified that tragedy was contagious.

"I was just so—" She breaks off.

I push a glass of water toward her.

"I needed to talk to someone who had known her. Someone who had played with her and cared. A friend, anyone. But I didn't feel like there was anyone. So I called Whirlpool and told them the washing machine was broken again. And then I had to break the washing machine."

"You *broke* the washing machine?"

She nods.

"How?"

She doesn't remember exactly ("Perhaps a stick?").

"But he came. And I told him—all about what had happened."

This hangs in the air for a few seconds until I can ask: "What did you tell him?"

"I told him how the ambulance came first, within twenty minutes, and how two paramedics tried to restart her heart." Her voice cracks. I drink the glass of water too fast and feel sick. She goes on: "Then the police came. Then the doctor. Then the coroner. Then the priest. Then the mortician. I told him how you came into the nursery while I was sitting on the floor holding Sophie before she was taken away. And how you said, 'Sophie's asleep, put her in her crib.' And how I wanted to explain what had happened but I didn't have the words. And how, after hours of being questioned, I wrapped her up in her blanket and had to hand her over to the police officer. And then your grandma came and took us to her house for roast chicken. And your dad went back to work the following week."

Her voice wavers. She tells me that the washing machine "broke" after my sister's funeral, too.

"The Whirlpool man came back and we talked again. I told him how I wanted to see Sophie after the autopsy but everyone told me I'd better not—that the autopsy would have been . . . *very thorough*. And about how she'd had a tiny white coffin padded with swan down that

I couldn't afford. Swan down!" She shakes her head. "I remember I asked the funeral director, 'Is this really necessary?' and he told me: 'You want the best for your little girl, don't you?'"

Now we're both crying.

I'm also filled with rage at a funeral director who "upsold" my grieving twenty-seven-year-old mother *swan feathers* when she was at her most vulnerable. How could someone do that? How could she bear it? How *did* she bear it? I'm not sure. But she did. She *does*. She is braver than either of us knows.

She tells me all about the "death-admin." About all the paperwork: a pink document from the coroner; a green one from the registrar before the burial. About how four grown men carried my sister's coffin in church. "Which was ridiculous as I could just as easily have carried her myself. I *wanted* to carry her myself."

My mother tells me about the funeral service, and then how everyone ate sandwiches afterward, before being expected to "move on."

"But I couldn't."

No.

"And the worst question a stranger could ask was, 'How many children do you have?' Because they never wanted to hear: 'I've got two children but one died.'"

I feel as though someone has kicked me in the stomach. My mother will later tell me that this is a sensation she has regularly. I'm reminded of something Julia Samuel said, about how the person has died but the relationship continues. About how we're not only mourning the person who's gone: we're mourning the death of the future we expected. Life has been interrupted. And we're suddenly awarded membership of a club we never wanted to be a part of. And my mother knew that she was going to be in this club forever.

"And you told all this," I ask, when words will come, "to the *Whirlpool man*?"

She nods.

All vegetable preparation is abandoned and we cry until our heads throb and our brains shrink. And this, I think, is strangely good for us.

My mother needed someone to talk to. She needed a buddy. She got the Whirlpool man ("Thank goodness for the service contract"). We all have our version of the Whirlpool man—someone who saves us, when we're at our lowest ebb and need it most.

Many of us were brought up not to talk to strangers. But perhaps we should. More. In 2013 the psychologist Elizabeth Dunn ran a study that found that even minimal social interactions led to a greater sense of belonging and positive effect.[3] Connecting with strangers makes us feel "understood" by humanity. Included in it, even. And we confide in strangers more than we might think—from barbers like Ben May to the person we're stuck next to on an airplane. Studies by the Harvard sociologist Mario Small have found that we regularly discuss important matters with people we're not intimate with, either because we think they might be knowledgeable or because they're just *there*.[4] They're also a blank canvas: an unknown, unbiased pair of ears. I've been that stranger, too, at times. A recipient of secrets and outpourings from strangers who wanted to talk but didn't want to tell their nearest and dearest for fear that it would worry them, or that they might tell someone else, or that there would be consequences. But we all need to be heard, by someone.

The Buddy System is the ideal. But if ours isn't quite up and running yet, a kind stranger can help. And should. If you haven't stumbled across your Whirlpool man yet, you will. Someone with a big heart who treats yours kindly, without a second thought, because it's the right thing to do. If he's out there, still, I thank my mom's Whirlpool man from the bottom of my shredded heart, for being there. For listening.

Part III
Stuff to Do When You're Sad

This is me. My part of the story is nearly done—we're almost up to speed. So here's what we can all do when we're sad to feel better. Not "not-sad" but *better*. The goal is to feel "good-sad," "helpful-sad" and "change-prompting sad." And to give ourselves a fighting chance of stopping normal, healthy sadness from tipping over into something more serious.

From culture as cure to thermostat wars and cold-water plunges. How spending time in nature helps us to be sad, well, and exploring balance versus burn-out. The smartphone paradox; why physical resilience can lead to inner strength; and the importance of "active acceptance." Plus why we all need to do something for someone else, identify as activists and step up as allies. Now.

Introducing a seal named Derek, skinny dippers, SUP and some seriously good news about hummus. Featuring Mozart; Jack Johnson; James Wallman; Svend Brinkmann; Frederick Douglass; Mary Wollstonecraft; Jo from *Little Women*; weather psychologist Trevor Harley; physiotherapist Dr. Brendon Stubbs; Alex Soojung-Kim Pang on the importance of rest; Joshua Becker on minimalism; Professor Felice Jacka on food and mood; Ella Mills on balance; and Nompumelelo Mungi Ngomane on how to "show up."

17

Take Your Culture Vitamins

I DRIVE MY MOTHER to the airport the next morning. Sad to see
her go and just . . . *sad*. It hurt, to talk to her like that. I know it
hurt her, too. But it's good pain. Like a muscle that aches from being
well used. Rain falls, faster than my windshield wipers can flick it
away, so it takes a while to realize my blurred vision is also a result of
tears. And then I'm home.

I hear my family before I see them, from the street in fact. This
isn't wholly unusual. Inside, there appears to have been some kind of
mud-powered mutiny, with earth-caked boots flung about and classical music blaring at full volume. This last part, at least, isn't the norm
in our home.

What's happened is that someone has been fiddling with the radio.
By the smears of peanut butter down the side, my money is on boy-
twin. Whoever the culprit is, they have inadvertently wiped all our
preset radio stations and replaced them with Classic FM.

What? Why? How? Even *I* don't know how to change the presets.

Boy-twin stomps in, licking remnants of nut butter from one hand like a bear cleaning a paw. He nods toward the radio: "I press buttons."

"Right. Yes, so I hear. *Which* button did you press?"

"I press . . ." He ruminates, then beams: "ALL the buttons!"

The Scandi chic minimalist radio that T bought for the kitchen is far too cool to have anything so useful as *markings* on each button, so I too start pressing "all the buttons" at random, when T comes in.

He winces and recoils slightly at the din: "Is this Stevie Wonder?"

"No." I squint at the display to see what's playing now: "It's Puccini . . ."

"Oh."

In case it needs clarifying, neither of us come from big classical music families (did you guess?). I once attended a performance of Mahler ("concert"? "show"?) in the hope of impressing a boy (to be fair, it was Mahler—he was impressed. I was emotionally pummeled). Other than that, I haven't voluntarily engaged with classical music since school. But now, gathering around the radio like an advertisement for the early days of "the wireless," I can't deny there's a sort of sonic solace to the soaring crescendos of Puccini, or the melancholy of Benjamin Britten's "Concord" from *Gloriana* that comes next and soothes our jangled nerves.

Peanut butter-covered surfaces (including children) get wiped. The mop comes out to tackle the mud. The kettle goes on. And then we sit. And listen. And *feel* something. Not exactly "relaxed" but . . . connected.

Classical music has long been proven to have psychological benefits and the "Mozart Effect" was established in the early 1990s.[1] This is the idea that listening to Mozart significantly increases our spatial reasoning skills. What's less widely reported is the fact that this effect only lasts for ten to fifteen minutes.

But there's more. I know from previous research that music therapy has been shown to reduce psychological stress, depression and anxiety among pregnant women, according to a study from the Kaohsiung Medical University in Taiwan.[2] Another study, from Juntendo University Hospital in Tokyo, Japan, played mice recovering from a heart transplant either Verdi's *La Traviata*, Mozart, or Enya.[3] They found that mice who were played classical music during recovery lived almost four times longer.[4]

Playing sad music when we're feeling low produces a heightened emotional effect that can foster a sense of belonging, give us identity and even help us heal, according to science. Studies show that when we're depressed, we're more inclined to seek out sad music. Researchers from the University of South Florida played both depressed and non-depressed people excerpts of "sad" music ("Adagio for Strings" by Samuel Barber and "Rakavot" by Avi Balili), as well as happy and neutral music. Researchers found that depressed participants were more likely to choose the sad music, because it was relaxing, calming or soothing.[5] Another study from the University of Limerick showed that nondepressed people also prefer sad music when blue because sad music can "act like a supportive friend" and trigger bittersweet memories[6] (like *saudade* in Chapter 12). Crucially, sad music also functions as an "acceptable" distraction, allowing us to escape the silence and feels somehow more appropriate when we're down. As much as I love Whitesnake and Van Halen (a lot, thanks for asking), upbeat music can feel unthinkable during times of sadness. But sad music can break the silence while also acting as a companion to our own low mood or troubled state. It can encompass wider human suffering, give us a sense of perspective and make us feel as though we're not alone.

Many of us come to rely on this and those who listen to music for three or more hours a day claim it is more essential to them than coffee, sex or TV. Thirty-eight percent of people reported feeling stress-free while listening to music, despite only 5 percent of people saying they have a "stress-free" life, according to one Sonos study.[7] Some argue that if classical music doesn't hit the spot, pop can help, too. Certain songs have extra emotional weight for personal reasons, while others are born out of such searing tragedy that they can't help but have an effect on us. Like Eric Clapton's "Tears in Heaven," written after his four-year-old son Conor died. Or the late George Michael's "Jesus to a Child"—a tribute to his partner Anselmo Feleppa who died from an AIDS-related brain hemorrhage. George Michael couldn't write music for eighteen months after Feleppa died, but then wrote "Jesus to a Child" in just over an hour. This helped him mourn his loss and he would dedicate the song to Feleppa whenever he performed it live. Then there's the Temptations's "I Wish It Would Rain," written by twenty-three-year-old Rodger Penzabene whose wife had just left him. He wants to cry, to "ease the pain," but men "ain't supposed to cry," so he wishes it would start raining to hide his tears. But it doesn't. He can't "ease the pain" and Penzabene ended up taking his own life on New Year's Eve 1967, a week after the single's release.

Music can hit us in the solar plexus. It can make our knees buckle, taking us back in time to remember and process emotions that still haunt us. Certain songs reliably make me stop in my tracks. "Save Tonight" by Eagle-Eye Cherry catches my breath, every time (unsuitable boyfriend + first love = a collision that knocked me for six). "Love Me or Leave Me" by Nina Simone, ditto (the tall guy, who took the lyrics literally). My mother turns to the music of her youth when she feels sad, she tells me: "Simon and Garfunkel and Janis Ian, mostly." It is telling that I have never heard a Janis Ian track in my life, despite spending my first eighteen years and a good chunk of my late twenties living with my mother. She did her "sad" in private. We all did our "sad" in private. But I feel comforted to know that at least she had songs to help handle the hurt.

Even more low-key music can help when we're struggling, according to Mikael Odder Nielsen, manager of the *kulturvitaminer*, or "culture vitamin" program near where I live in Jutland, Denmark. Nielsen offers people suffering from stress, anxiety or depression the opportunity to go on a culture crash course. "We use playlists developed by music therapists to give your brain a break, which in turn allows your body to take a break," says Nielsen, explaining that it's music to "reduce arousal": "music that's predictable, *boring* even."

Such as? He thinks about this before responding.

"Jack Johnson."

Huh . . . Popping on "Banana Pancakes" isn't something I'd naturally have gravitated toward when feeling blue, but I try it later and strangely it works (ish). I now think of JJ as the audio equivalent of a mindfulness coloring book. I tell Nielsen I may be a convert and he tells me about some other strands of the *kulturvitaminer* program. Partly funded by the Danish Health Authority, municipalities set up programs encouraging cultural participation for the unemployed or those on long-term sick leave.

"We wanted to see if we could improve people's mental health, reduce social isolation and help them get back into the labor market via culture," explains Nielsen. Qualifying residents were invited on two to three cultural excursions a week, for ten weeks to see if it made them feel better. And it did. I speak to Jonas, who suffers from anxiety and was fearful of social interactions prior to starting the course. But the program, he tells me, changed his life.

"It was an activity that would get me out of the house, where I was treated as 'normal.' And I am not my anxiety: I'm me. So the course helped me feel like 'me' again."

A former preschool teacher called Evy tells me how she suffered from stress and chronic insomnia for six years. "Before I went down with stress, I would often go to concerts and museums," says Evy, "but then I stopped. Nothing made me happy or even made sense anymore." This strikes a chord.

"If you're depressed, culture is often the first thing you don't bother with," Nielsen explains, "you're too preoccupied with getting through the day. My role is to get them used to this world again—or even introduce it for the first time."

Structured, *ritualized* meet-ups and planned outings also get people back into culture and its benefits by stealth: we may not feel like going out and socializing and engaging with live art when it's raining and we've got Netflix at home. But if it's part of a course or we're somehow "committed"—either financially or socially, by promising we'll meet someone there—we're more likely to show up. And the "stealth" element means that Nielsen has found the course especially beneficial for men who may be more reluctant to express their negative emotions and vulnerabilities otherwise (thanks, social conditioning). It's the Buddy System *plus* rituals—or culture on a timetable (and who doesn't love a timetable? Not me: I *love* a timetable).

There are eight strands to the Aalborg culture program, starting with group singing, proven to help forge social connections and bond large groups.[8] The course includes trips to the city archive to learn about local history and "foster a sense of belonging and local pride," says Nielsen. Participants also go on theater excursions, visit art galleries and take part in creative workshops, proven to help develop resilience.[9] They even take in performances by Aalborg Symphony Orchestra ("Very moving," says Nielsen, "there are often tears."). This is smart, since researchers from the Royal College of Music have found that hearing live music reduces stress.[10]

Of course, lockdown threw a wrench in the works here. Many galleries, theaters and concert spaces worked to make their art available online but COVID-19 still poses an existential threat to the performing arts. The impact of the pandemic on the arts economy and artists will be felt for years to come—something we can all feel sad about. But we can also acknowledge and appreciate the value of culture and how much it means to us—how much it's worth hanging on to.[11]

Most of us will remember looking at a piece of art that moved us. Just last year, T had to remind me to "breathe" while looking at the extraordinary light captured in the Skagen Paintings—by a group of Scandinavian artists who gathered in the village of Skagen, the northernmost part of Denmark, from the late 1870s. Mid gallery visit, I came over all peculiar and had to sit down. There's a name for this, apparently. Stendhal or "Florence" syndrome is a condition whereby we become dizzy, faint and overcome in the presence of beauty or impressive art. The syndrome is named after the nineteenth-century French author Stendhal,[12] who, on visiting Florence, was so awed by the beauty and art all around him that he "walked with the fear of falling." Sometimes the only response to any kind of overwhelming emotion is to fall down and stay there (I've never been to Florence, but God help me when I do).

Culture as "cure" isn't a new idea. Research shows that "art on prescription" is valued by referrers and participants alike[13] and is remarkably cost-effective[14] with a reduction of visits to the doctor and participants gaining transferrable skills that can help with staying healthy and in work.[15] Sweden leads the field in terms of arts on prescription in Scandinavia, but Australia has had a National Arts and Health Framework to promote integration of arts and health since 2013.[16]

There's now a substantial body of evidence showing how the arts can mitigate some of the negative effects of social disadvantage and a 2015 review found that the community cultural referral schemes led to increases in self-esteem, a greater sense of empowerment and reductions in anxiety and depression.[17]

"We can see that it works for many people," says Anita Jensen, post-doc researcher from the Department of Communication and Psychology at Aalborg University who I speak to about the Danish scheme. "It's relatively inexpensive—with no known negative side effects." (Other than the Florence syndrome.)

Another popular strand of the culture vitamin course is "shared reading," where adults are encouraged to snuggle up under blankets in a dimly lit breakout room in the library while a librarian reads aloud to them for two hours at a time. Which sounds a lot like heaven and is very much what I'd like for every birthday and Christmas from now on. Most of us haven't had books read to us since childhood—audiobooks aside. I say "most" since apparently some people end up with life partners so adoringly romantic that they insist on reading to their beloved— poetry, philosophy, great works of literature, Harlequin romances, you name it. To these folk, I extend felicitations. To the rest of us, I say: Nope, me neither. I imagine that being read to is an intimate, nurturing experience that can be pretty emotional, too. As Evy tells me: "I spent so much of my life reading to others in my job as a teacher, but this time I needed help—and I felt . . . *taken care of*. It was very powerful."

When we're depressed or suffering from anxiety, many of us find reading difficult. I can't concentrate or make any sense of the words on a page when I'm depressed. The silence necessary for reading also feels intolerable. But being read to—via audiobooks in my case—is something different. It's like being taken gently by the hand and taken on an adventure (in a nice way). I listen walking home from dropping the kids off. I listen in the car, or while shopping for groceries, or while cooking. I'll listen at night, too, if I can't sleep. With an audiobook, nothing is expected of us other than to listen. And it is, unfailingly, enriching.

Now in Denmark, the Aalborg council has chosen to prioritize cultural vitamins and run the program on an ongoing basis. And another Aalborg local is campaigning for culture as cure throughout Denmark and much of Europe. The Stoic philosopher Seneca recommended that we read poetry, gaze at green objects and play the lyre in order to live a good life. Nietzsche believed that art was the medium through which we are unified as human beings and we could all experience catharsis through tragedy. Today the Danish psychologist and philosopher Svend Brinkmann is getting more specific, prescribing novels as a north star for modern life.

Brinkmann is fast becoming a national treasure in Denmark. As a "lowly college professor" he operated under the radar until recently, writing academic texts "that about eleven people read," he tells me when we speak. Then in 2014 he published *Stand Firm*, a satire on traditional self-help that became a bestseller (*Stå fast*).[18] Now Brinkmann's a regular on Danish TV and radio and often appears internationally. He can seem intimidating at first glance, thanks to strong views, a vast intellect and a series of particularly fierce-looking publicity photographs. Fortunately, I'm feeling brave when we speak and he proves to be far more congenial than I'd feared ("I don't know why people think I'm so scary!" he tells me. "It's not how I see myself at all—I'm the guy who will lie awake at night worrying about everything!"). Brinkmann's a busy man and since he published *Stand Firm*, he's become a leading advocate for the power of the novel as the ultimate personal-development read.

Novels matter, Brinkmann argues, not just because they are well-written pieces of art, but because they explore what it is to be human. A good novel examines what life is all about and helps us maintain a sense of perspective with regard to our own existence. Novelists aren't restricted to speaking in just one voice, but can use multiple voices that may even contradict each other. Thanks to their "polyphonic nature," novels can teach us to understand and appreciate other points of views. And this is, quite literally, "mind improving" says Brinkmann.

18

Read All About It

B RAIN SCANS SHOW that when we're immersed in a book, we
mentally rehearse the activities, sights and sounds of a story,
stimulating neural pathways.[1] Reading has also been shown to boost
empathy levels and help us to connect.[2] As the philosopher Alain de
Botton writes in *A Velocity of Being*:[3] "We wouldn't need books quite
so much if everyone around us understood us well."

But they don't. So we do.

If we're depressed or suffering from anxiety and can't read, audiobooks can help. Stories can help keep "normal sad" from tipping over
into something more serious and we can use books as part of our
mental health maintenance. Especially, according to Brinkmann, fiction.

Novels help us grapple with moral conundrums and the "what if's"
of this world. They make us question our own behaviors and convictions.
And they can do this even when they were written long before our time.
Aldous Huxley's 1932 novel *Brave New World* described a future where
emotional pain is eradicated through the use of a wonder drug, Soma.

It was written as a dystopia. Only today, Huxley's description of a voluntarily emotionally sterile world doesn't seem *too* far-fetched. We all seek pleasure and avoid pain. In Freudian psychoanalysis, "the pleasure principle" is all about the instinctive seeking of pleasure and avoidance of pain to satisfy biological and psychological needs. And our technological advances mean that a brave new pain-free world isn't beyond the realms of possibility. In case I haven't been clear enough already: this would NOT be advantageous. So Huxley's *Brave New World* is a brilliant example of a novel helping us to consider, empathize, predict and ponder. Brinkmann's other recommendations for novelists to help us get our "polyphonic" on include Charles Dickens, Vladimir Nabokov, Haruki Murakami, Miguel de Cervantes, Michel Houellebecq, Cormac McCarthy and Karl Ove Knausgård. "Novels," he says, "teach us to 'stand firm,' by helping us find an external meaning or perspective on life upon which to do it."

But Brinkmann isn't a fan of nonfiction. In fact, he'd rather we all shut the books or step away from our Kindles. In *Stand Firm* (a nonfiction book—and yes, he notes the irony), he argues that if we read biographies and self-help literature we are "presented with the idea of the self as the inner and one true focal point for life" and "offered a positive and optimistic story of development in whose glory you are invited to bathe." This, he says, reinforces the idea that life is something we control. And as a heartily secular Danish psychologist with a background in philosophy, Brinkmann believes that it most certainly is not. It's fair to say that if there were a sliding scale of pompom-waving "You can do anything you want as long as you want it enough!" types, with Tony Robbins at one end, Brinkmann would be at the other end. Possibly outside in the parking lot reading Nabokov. Biographies and self-help books: *no*, says Brinkmann. Novels: *yes*, since novels enable us to understand our existence as complex and unmanageable.

I'm conflicted here. I'm all for Team Novel (Should we get T-shirts made? I'll get T-shirts made). But I've always found biographies wildly compelling and remarkably reassuring. As the Irish writer Colm Tóibín puts it, "To cheer yourself up, read biographies of writers who went insane."[4] The biographies I read don't tend to tie "life" up in a neat little bow. Quite the reverse in fact: they reassure me that everyone is fallible and that other people have been through the same challenges we're grappling with. When I worry about juggling parenting with writing, I remember J.G. Ballard's autobiography[5] and how *Empire of the Sun* writer raised three children as a single father following the sudden death of his wife. His youngest daughter, Bea, was only five at the time. It must have been phenomenally painful for all of them. I try to imagine his daily routine; how he talked to his children; how he could parent and then "switch off" to write. I wonder how he managed it all—just as I do when reading about any life that isn't my own. This makes me think and empathize and feel connected to the universality of the human experience.

There are biographies about people who win wars and discover things, and then there are documentations of lives that often go untold. The domestic realm wasn't considered worthy of being set down in print form for the longest time, and anything women read or were concerned with was somehow considered "lesser." But I can't help thinking biography goes some way to redress this.

Louisa May Alcott's *Little Women* made me cry as a child when Jo passed up love with the hot young Laurie and ended up with the stupid old professor. But since reading more about Alcott as an adult, Jo's decision has taken on another level of significance. It turns out that Alcott, like Jo, wrote Gothic thrillers. While Jo, wearing the "moral spectacles" of the stupid old professor, threw her stories in the fire, Alcott published hers and made a fortune. She also wrote *Little Women* in ten weeks (#writergoals) and earned financial independence with a book that hasn't been out of print since it was first published in 1868.

Whereas the March sisters were all obliged to follow traditional "womanly roles" and even Jo got married (to the stupid old professor), Alcott never wed and wrote some pretty racy heroines in later life including an actress who impersonated a governess to seduce the master of the house.[6] Alcott was able to support her whole family on her impressive earnings. This was fortunate, since her father, Amos Bronson Alcott, viewed breadwinning as inconsistent with his transcendental idealism (don't we all?) and his drive to create a utopian society (eye roll . . .). Learning about *Little Women*'s author, her drive, passions and motivations only enhances my enjoyment of her work.

I first read Mary Wollstonecraft's *A Vindication of the Rights of Woman* as a cargo-pants-wearing beer-drinking student in the late 1990s. I found feminism's original manifesto both blisteringly angry and utterly inspiring. It was a torpedo in paperback form that made me want to do something—anything—to be *useful*. Learning more about Wollstonecraft's biography later on doesn't diminish this, it only increases my wonder at her work.

Wollstonecraft was the second child in a family of seven with a violent father who, one day, for no apparent reason, hanged the family dog. He sexually abused his wife for years and as a teenager, Mary Wollstonecraft set up camp outside her mother's door night after night in the hope of protecting her. She couldn't stop her father, but every night she tried. She was brave as well as brilliant. But she wasn't a saint.

Wollstonecraft had many attributes but "sharing" wasn't one of them. She once wrote to her best friend, Jane, that she'd really rather prefer it if Jane wouldn't go hanging out with other girls, saying: "I must have the first place or none."[7] On another occasion she wrote to poor old Jane: "I shall take it as a particular favor if you will call this morning, and be assured that however more deserving Miss R . . . [a "rival friend"] may be of your favor, she cannot love you better than your humble servant Mary Wollstonecraft. P.S. I keep your letters as a Memorial that you once loved me, but it will be of no consequence to keep mine as you have no regard for the writer."

Ouch.

Knowing more about Wollstonecraft's life helps us understand that we can be wonderful and revolutionary and kind of a pain in the ass, all at the same time. We can be fallible—human, even. And for women, whose stories have been silenced for hundreds of years, this feels important. The rise of feminism in the eighteenth century was accompanied by increasing diagnoses of "hysteria" to shut us up again. By the nineteenth century, there were commonly held beliefs that overeducating women might break our wombs (I paraphrase) and we'd all succumb to "anorexia Scholastica"—an incurable "disease" where women apparently experienced headaches, neuroses, epilepsy, severe weight loss, loss of "morality" and even comas. Which all seems a rather harsh punishment for simply liking books and makes me want to defend our right to read what we like, when we like. Biographies, this way, may well be a feminist issue. They're also an inclusivity issue.

I think of all the groups that have been excluded over the years—from LGBTQ+ figures to BIPOC voices—and it seems clear that representation is a battle we're still fighting. Biographies give a voice to people who may not have been heard before. Dr. Maya Angelou's memoir *I Know Why the Caged Bird Sings* is a classic example. Or *Narrative of the Life of Frederick Douglass, an American Slave*, the 1845 memoir by Douglass who was born into slavery in Maryland, and went on to become one of the nineteenth century's most prominent abolitionists. After the Civil War and the abolition of slavery, he campaigned for equal rights for African Americans, arguing against Lincoln who had wanted freed slaves to leave America. "We were born here," said Douglass, "and here we will remain." Frederick Douglass is a man whose legacy I knew little about until his autobiography was recommended to me in 2018. So, you know: important. I could go on (I will: Trevor Noah's *Born a Crime*,[8] about growing up in South Africa the son of a white Swiss father and a Black Xhosa mother, at a time when such a union was punishable by five years in prison, is astonishing). Suffice to say: biographies feel crucial to our understanding of the world.

I put this to Brinkmann who concedes that the anti-biographies part of his book was the most controversial ("That was the only chapter my editor pushed back against") but he explains that he had a particular genre of biography in mind. "It was the hero genre, with a hero who conquered the world at its core," says Brinkmann.

And this isn't good?

"No!" he tells me. "Because this is not true to most people's experiences—and we can only feel inadequate when we read these stories. Of course we need all sorts of voices. I'm actually a big fan of auto fiction [fictionalized autobiography]. It's the honesty of representing life as it really is. It's more complex and . . . *non-shiny*."

I understand his argument. Brinkmann is a disciple of the Stoics, who suggest we shouldn't navel-gaze too much and that we should try to take the "self" out of the equation wherever possible. Brinkmann is also Danish, where a code of conduct called "Jante's Law"[9] decrees that no one is to think they're special and the collective is prized above the individual (see: Nordic democratic socialism/eye-wateringly high taxes). Expressions of individuality and personal success aren't encouraged in Nordic culture traditionally. So it stands to reason that the individualistic approach of a book all about one person, *their* perspective, *their* trials and tribulations, may hold less of an appeal. Especially if that person is a successful, shiny, "look at me!" type, attempting to extrapolate worldly wisdom without the input of other people.

But there's another mode of thought, too. One that puts heroes front and center of the story.

The author and trend forecaster James Wallman, of *Stuffocation* fame,[10] recommends that we *all* think of ourselves as the heroes of our own journey. Rather than feeling restricted by the idea of the archetypal hero, he wants to reframe the "hero" narrative so that it can apply to all of us. I meet Wallman in London and we drink tea one rainy Saturday while he explains about the hero's journey and how to live by it.

"The hero's journey" narrative decrees that pretty much every story will follow the same narrative arc or basic plot. The "hero" will experience a call to adventure before having to cross the threshold of no return, and face tests, allies and enemies. There will inevitably be an ordeal where the hero experiences a major hurdle, before finally reaping the reward(s) and making the journey back to ordinary life. There are seventeen stages to the original hero's journey narrative structure, as observed and recorded by the mythologist Joseph Campbell in his 1949 classic, *The Hero with a Thousand Faces*.[11]

Wallman tells me about the Australian psychologist Clive Williams who read *The Hero with a Thousand Faces*, and began to wonder whether the story structure of the archetypal hero might be applied to real life, too. The more Williams thought about this, the more he noticed the "hero's journey" stages happening in his own life. He found this a helpful way to reframe the challenges he faced—so when obstacles or people got in his way, instead of feeling despair, he'd view them as a necessary part of his "journey": facing tests, allies and enemies. This led him to the idea that the hero's journey could be used as a "mud map"—or roughly drawn path—for living.[12]

"When I came across Williams's work, I had one of those Eureka moments," says Wallman, who writes about this in his latest book *Time and How to Spend It*.[13] "The more I've thought about it, the more I agreed: the hero's journey is an incredibly useful approach." By viewing setbacks and difficult times in our lives as essential stages in our own personal "narrative arc," Wallman believes we can handle them better and feel less demoralized when they rear their ugly heads. "It just makes sense to me," he says.

I mention Brinkmann and the Nordic approach, whereby we're not to think we're "special" or the star of any show ("even our own").

Aren't we all too self-obsessed already without turning ourselves into the hero of our own journey? Shouldn't we all be trying to be more Stoic?

Wallman thinks not.

"Stoics are *idiots*! What I mean is: I don't agree with their point of view. Of *course* it's all about us: we see the world *literally* through our own eyes, so it couldn't be any other way," he says. "It's stupid to think otherwise—a denial of the existence you have," he says, adding: "We are nothing more than farts in the wind, so we have to take control. Human beings need stories and we need to be at the center of our stories, because what's the alternative? You can't be a bit part in your own life." Wallman is resolute on this: "If you're awkward calling yourself the 'hero,' try 'protagonist.' But it has to be about you. If you're not the hero, you lose your identity. When we say people have 'lost the plot,' we mean that they've lost where they are in their own story." There is, surprisingly, some science behind this.

Studies show that people who tend to see their life as a story of personal growth—as a journey—have higher levels of eudaemonic well-being, the kind that's about a good life, rather than feeling jazz-hands happy all the time. I'm reminded of the childhood bereavement psychotherapist Ross Cormack, who begins working with bereaved families by constructing the story of what's happened with them—a necessary narrative that has a beginning, a middle and an end. People who frame difficult times as "transformative experiences" and view suffering as a path to gaining new insights also fare better psychologically.[14] But the downsides of such standard narratives have also been documented, resulting in stigma for those who don't follow the blue-print, or unrealistic expectations of happiness for those who do. Once we get to the part of a story where it's supposed to be happy ever after, and it's not, it doesn't feel great (see Chapter 9 and the fallacy of arrival).

Wallman is convinced that the pros outweigh the cons: "It certainly helped me," he says, "and there have been some seriously stressful times." He gave up his job when his wife was also off work having their children, to write and self-publish his first book. "There were financial challenges," he says, "and it was tough on my wife and on our relationship. I thought, How the fuck am I going to make this work?" It was, he says, "proper hero's journey stuff: thinking about allies and mentors and enemies."

I ask about enemies in these divided times. Can it really be helpful to "other" and appoint our own nemeses?

"They don't have to be people," he points out, "it could be anything from the wind against you." He gestures outside to where Storm Dennis is currently tearing apart central London. "It can be the enemy *inside* us," he goes on, "it does not have to be other people. But we do need stories. And we should be at the center of ours."

I think about this for a few days. We are narrative creatures—we need stories to make sense of our lives. And we have a natural impulse to make sense of what we experience. I have, on more than one occa-sion, frantically Googled "Halloween 1982" in the hope that something momentous happened on the day my sister died. Some sort of meaning or logic as to why we lost her, then. Some clue that will unravel the mysteries of the universe. But I am disappointed, every time.

I broaden out my search to the events of that week and discover that Pope John Paul II went to Spain and Arnold Schwarzenegger was on the cover of *Life* magazine.

Pope John Paul II and the Kindergarten Cop?

Real life *can* just be random.

In these uncertain, polarized times, I resolve to sit firm in the grey area here, embracing biography *and* stories. Making sense of what I can, when I can. I read, a lot. Except when I'm really sad and the words dance around or I'm too tired to turn a page, when I plug in an audiobook. And I feel connected—to the world around me, past and present, by invisible golden threads. I feel exorcised, understood and more empathetic, all at once.

My culture vitamins consist of novels and biographies. And I'm fine with that. Both genres depict all the messy trouble of human life in a way that is consoling and helps develop resilience. With the polyphonic novels that Brinkmann recommends, we can have a trial run at empathizing all the hurts we've yet to personally experience. With biographies that go beyond the "straight white middle-class man achieves further untold riches" rigmarole, we learn more about other people's lives and begin to see our own as a series of inevitable trials and challenges. Books may be our only access to these obstacles before we experience them. Stories may be our sole benchmark for specific sadnesses, before we collide into them at 3 p.m. on an idle Thursday. So they're worth investing in.

The comedian Robin Ince also co-hosts a podcast, *Book Shambles*, with fellow comedian Josie Long. Books have always been important to him, and when we speak he's standing in a room stacked waist-high with books. "I do *love* books," he says. "It's the nearest I come to being a hoarder. Every time I look at a new book I think, There are going to be ideas in that! You get lost in someone else's narrative, it's really . . . useful," he says. And he's right.

We often think that our situation is unique and no one knows what we're experiencing. But when we read accounts of other people's struggles and perspectives, the illusion of our own "sadness" being freakishly rare fades away. And far from having lost the plot, this is A Good Thing.

There's also a value in the sheer beauty of art that can make us notice our world anew. The last time I felt properly eviscerated by sadness, as if I'd lost my skin and had to decide whether to allow myself to be exposed, or whether to calcify and armor up, I chose art. And I began to notice the things around me. To be lifted by the unexpected joy of a lovely turn of phrase. Moved to remark upon a really pleasing spiderweb. Enjoying a truly excellent cup of tea. Brought to tears of gratitude by a bright but blustery morning. Having a snot-cry in an art gallery on a particularly hormonal Tuesday, walking around and thinking about all the lives pictured therein. All the people. All of us.

Being able to sit with sadness isn't a weakness: it's a strength. By giving sadness the space and time it needs, we're likely to spend less space and time in it, overall. None of us need to wallow. But just letting sadness *be*—letting it run its course—can help.

The arts can meet the existential challenges of life and find meaning in suffering, loss and death. Sometimes. Culture has the capacity to make sense of the apparent randomness, bringing in order and a new way of approaching life. Sometimes. Taking our culture vitamins can help us access deeper and more nuanced thoughts and feelings than the daily grind of Western society normally allows for. So after speaking to Nielsen and Brinkmann and Wallman and listening to the Temptations on repeat while weeping for Jo March, teenage Mary Wollstonecraft and George Michael, I take my culture vitamins. On days when I feel too fragile even for this, a good drama or movie works well, too. Anything to give me another perspective that will enrich my own.

I convalesce from a long, dark winter by reading, writing, listening and looking around me with fresh eyes. I lick my wounds and recover, until I feel stronger. Until I can face the world and get out there.

19

Get Out (& Get Active)

BLINKING INTO THE light, the world looks different. The birds are back, after a long, dark winter; green buds tip the ends of the rose bush outside our house; and . . . what is this vast, luminous orb putting in an appearance? Could it be? Could it really be . . . the sun? Almost instantly, a large slate-colored cloud floats across the sky and obscures my view, but I remain elated at this merest crack of sunlight—nature's molly—that makes me feel as though surely, (surely?) everything's going to be okay. The rest of the day, I am dancing.

Extraordinary, I think, can it genuinely be that the weather has such an impact on mood? This seems a little too neat. Is my brain really so basic?

"Yes," is the short answer, followed swiftly by "and so is everyone else's."

Professor Trevor Harley is one of the world's few psychometeorologists—working at the intersection of psychology and the weather—and he's also the author of *The Psychology of Weather*.[1]

I ask him what's going on in our heads and why I've spent the past twenty-four hours high on sunshine.

"Well," he begins, "the brain is a complex organ, so we don't know *why* exactly it has such an impact." (I'm getting used to the brain experts' by now familiar response of "*not sure*.") "But we have evolved to cope with an optimal temperature of around 70 degrees Fahrenheit," says Harley, "so anything too hot or too cold will have an effect on how we're feeling."

In fact, a 2017 study published in the journal *Nature Human Behaviour* found that 71 degrees Fahrenheit was the temperature that makes most of us more agreeable, emotionally stable and open to new experiences.[2] Researchers gauge "thermal comfort" by how many people will feel uncomfortable at any given temperature using the Predicted Percentage of Dissatisfied (PPD). But to calculate PPD, building managers use a formula from the 1960s and take into account the metabolic rate and inner thermostat of a 154-pound, forty-year-old male. I read a study from Maastricht University Medical Center confirming what I've long suspected: that most women have significantly lower metabolic rates than most men and like their air temperature 5.4 degrees Fahrenheit warmer.[3] This is especially evident in the workplace with thermostat wars and the need for an "office sweater" on the back of every woman's chair at every establishment I've ever worked in. There are also individual preferences. Mark Zuckerberg apparently keeps the thermostat at Facebook HQ set to a frigid 59 degrees Fahrenheit to focus the mind,[4] while President Obama kept his Oval Office so hot, an adviser joked to the *New York Times*: "You could grow orchids in there"[5] (he and I would get on like a house on fire/a house optimally heated).

There are, apparently, some universal truths no matter where we are in the world. "Wind is generally disliked," Harley tells me, "whereas snow is *very* popular. At least the expectation of it. But then it doesn't often come and so there's disappointment." Low mood is related to temperature and wind, I learn, although the variables that appear to have the greatest effect on mood are humidity and hours of sunshine. When the sun shines, we are nicer to each other or more "prosocial" as the psychologists put it. One study even found that we tip more generously when it's sunny.[6] But we don't like it too hot, or too sticky.

"When the weather is humid we experience discomfort, and this leads to irritability," says Harley. The link between heat, humidity and irascibility has been around for centuries ("For now, these hot days, is the mad blood stirring," says Benvolio in *Romeo and Juliet*). And violent heat and humidity have been catalysts for civil disturbance. This is a concept that's always surprised me. When it's hot and humid, I usually remain as stationary as possible, too enervated and sapped of energy to be bothered to do anything.

Who are these people with the vim to start a riot when it's muggy out?

Harley takes my point but tells me: "There's an inverted u-shaped curve—if the humidity is too bad we can't be bothered to get out of bed to go and have a fight, but under just the right levels of humidity, we can be irritated into action and snap." This isn't always the case (it can't have been too clammy during the Russian Revolution, for example). "But it's a trend we see quite often," says Harley.

"Then of course the greatest impact we see of weather on mood is in people who suffer from seasonal affective disorder or SAD, where lack of sunshine can lead to clinical depression," he says. It's not altogether clear why lack of sunlight makes us depressed, but one idea is that the amount of light affects the levels of melatonin and serotonin in the brain, and light stimulates activity in the hypothalamus. This is the part of the brain involved in regulating body functions, including sleep, appetite and, to some extent, mood. Incidence of SAD predictably varies depending on where we live—and it's especially prevalent in Scandinavia (lucky me!).

"Reverse SAD is less well known and less understood," Harley tells me. This is when we feel worse during the long, light days of summer and better during the short dark days of winter. "At the moment it's estimated that a tenth of all seasonal affective disorder cases are in fact reverse SAD," says Harley. The treatment for reverse SAD is the opposite of that for the "normal" variety: maximizing time spent in the dark, or at least away from sunshine and turning up the air conditioning. A scenario that currently seems so alien to both of us that we can't quite compute it.

The Danish winter reliably lasts from October until March and a study from the Ministry of Climate and Energy showed that there were only forty-four hours of sunlight in Denmark in November. That's just over ten hours a week—less than an hour and a half a day. The Danes even have a term for their own brand of SAD: *vinterdepression* or "winter depression" [*sad face, umbrella, thunderstorm emoji*].

It's not much better where Harley lives, in a dip between two hills, in Scotland. "In winter, the sun only appears over one hill at around 10 a.m., then disappears back behind the other at 2:30 p.m.," he tells me. To counter this, he uses a bright natural frequency light and tries to get outside a lot. "Not always possible, but that's the goal," he says, "despite the strong urge to sit at the back of a cave and hibernate in winter." At this point in our conversation, a poodle scrambles up onto Harley's lap. "Of course, this one helps." Harley tells me that the dog is called Beau. "He's been with me for two years now and I love him: I get oxytocin, I get exercise, and he gets me out of the house—I get to see some greenery. Maybe even a tree . . ."

Dogs have long been associated with encouraging physical activity and improving mood (T swears a man in his village growing up had a dog named Prozac, in honor of the hound's healing properties). Trees, too, have a long list of health-giving attributes—and we don't have to hug them to experience the benefits. An experiment led by

University of California, Irvine Professor Paul Piff found that participants who stared up at very tall trees for as little as one minute experienced increased awe and demonstrated more helpful behavior than participants who spent the same amount of time looking up at a high building.[7] Trees 1: urban planning 0.

There's a soothing feeling that we experience when we hang out in woodland, and in Japan, there's the recognized stress management activity of *shinrin-yoku* or "forest bathing." No water is required for forest bathing, it's simply spending time in a forest for the purpose of relaxation. Forest bathing has a surprising amount of science to back it up[8] and has been shown to reduce blood pressure, stress and anxiety,[9] thanks to an essential oil that boosts our immune system.[10] Admittedly this sounds far-fetched, but Professor Qing Li from Nippon Medical School in Tokyo has put forward a theory that trees emit phytoncides— a little like essential oils—to help protect trees from harmful microbes. When we breathe these phytoncides in, they prompt biological changes in us, too, and our bodies produce more of the natural killer cells that form our crucial first line of defense against viruses and tumors.[11]

The health benefits of spending time in nature have been felt intuitively forever, but now there's meta-analysis to unequivocally prove the impact of "green space" on body and mind.[12] Studies show that spending time in nature is good for our mental health, lowers our heart rate, lowers our blood pressure, reduces stress and can make us more resilient.[13] And it's this last part that I'm fascinated by—because spending time in nature is risky.

We may fall, or cut ourselves on brambles, or be stung by nettles. It may rain. Or we may be cold. We may have to bear countless other discomforts. But these make us stronger. Getting out and spending time in nature teaches us how to be sad, *well*—and helps shore up our mental health against the Big Stuff.

Researchers from Aarhus University in Denmark discovered that children who spend more time playing outside in green spaces are less likely to develop psychiatric disorders as adults.[14] Norwegian play researchers Ellen Sandseter and Leif Kennair studied children's risky play from an evolutionary perspective and confirmed that it had "antiphobic effects," helping us to extend our coping abilities incrementally and so empowering us to take on greater challenges. What's more, they say, "we may observe an increased neuroticism or psychopathology in society if children are hindered from partaking in age adequate risky play."[15]

For those born after 1994 (iGen or Gen Z) anxiety levels and depression have never been higher, says social psychologist Jonathan Haidt in the 2018 book he co-authored with Greg Lukianoff, *The Coddling of the American Mind*.[16] The 2018 American Psychological Association's Stress in America survey focused on the concerns of Americans aged fifteen to twenty-one and found that Gen Z is also significantly more likely (27 percent) than other generations, including millennials (15 percent) and Gen Xers (13 percent), to report their mental health as fair or poor. Nine out of ten Gen Zers (91 percent) said they have experienced at least one physical or emotional symptom because of stress, such as feeling depressed or sad (58 percent), or lacking interest, motivation or energy (55 percent), and the survey concluded that Gen Z is the least likely generation to report good or excellent mental health.[17]

The *Harvard Business Review* published a study by the mental health advocacy group Mind Share Partners in 2019 that found half of millennials said they'd left a job at least partly for mental health reasons. For Gen Z, the percentage rose to 75 percent—compared with just 20 percent among the general population.[18]

Previous generations lived through wars. But today's kids are less well equipped to cope and so are suffering more. Of course there are new pressures—like social media and the prospect of competing to enter an unstable workplace with far less job security than previous generations enjoyed. Still: *wars*. Today's young people are also indoors more, only spending half as much time playing outside as their parents did, according to studies.[19] This is significant. Correlation is not causation, but it can't help—and we coddle at our peril.

A 2013 review of stress research called "Understanding Resilience" claimed that: "Stress inoculation is a form of immunity against later stressors, much in the same way that vaccines induce immunity against disease."[20] If we don't immunize children with this kind of stress, we're storing up problems for them later. Spending time outdoors and free play, where children work out their own rules, learn to take risks and master minor dangers (poison ivy, big trees, perhaps a leaf fight), matter.

Since the 1970s, the distance a child is allowed to roam from home has shrunk significantly in the US.[21] A poll commissioned by the Children's Society found half of all adults questioned thinking that the earliest a child should be allowed out unsupervised was aged fourteen.[22] And if kids aren't even allowed to walk down the street by themselves, the chances of them exploring the natural world are even more remote. But we're missing a trick here, as Dr. William Bird, medical adviser to Natural England, puts it: "The outdoors can be seen as a great outpatient department whose therapeutic value is yet to be fully realized."[23]

Other countries do a little better. Forest schools in Scandinavia and Germany routinely turf kids out and let them fend for themselves. At a standard Danish day-care, children play outside, come snow or sleet, returning home with cuts, bruises and (only very occasionally) black eyes. My children are also regulars at the terrifying institution known as "family scouts." Essentially: spending three hours on a Sunday watching preschoolers whittle, wield axes, play with matches and generally juggle hazardous implements. It is at this august assembly that I first hear myself utter the now famous family catchphrase: "You can have a saw once you're four!" To a two-year-old.

The preschooler already has his own tool kit, complete with hammer, saw, screwdrivers (flat head and cross head), spirit level and measuring stick. The kid can't read or write, but he *can* build a boat from discarded plywood and kindle a half-decent campfire (oh, and I've since softened my stance on toddlers packing tools: I now let the two-year-olds saw *with supervision*. So far, all digits remain intact and they're learning about risk—a valuable life skill). Scandinavian kids run and jump and climb and fall and fail and get up again, in the fresh air, spending time in nature, for hours every day. Norwegians hold the gold standard for outdoor living among Scandinavians with *friluftsliv*—or free air life—a secular religion in Norway. During my last research trip there, I was alarmed to find Norwegians getting their *friluftsliv* in weather so cold that my eyelashes froze open. Now that's what I call commitment. The rest of us don't have to incur frostbite, but we should be getting out there, more than we probably are now. The benefits of green space are, by now, undeniable—and being by the blue is even better.

Spending time in and around water has been shown to improve our mood, reduce stress,[24] and even make us more eco-aware.[25] The closer we live to the coast, the healthier we are[26] and spending time by the water may help us sleep better. Researchers from Northwestern University found that people who fell asleep listening to water slept more deeply and boosted their memories[27] while a study from the UK's National Trust found that participants slept on average forty-seven minutes longer at night after a seaside hike.[28]

"Many of the processes are exactly the same as with green space—with added benefits," says Dr. Mathew White, an environmental psychologist with BlueHealth, a program researching the health and well-being benefits of blue space across eighteen countries. "We find people who visit the coast, for example, at least twice weekly tend to experience better general and mental health," agrees Dr. Lewis Elliott, also of BlueHealth. "Some of our research suggests around two hours a week is probably beneficial, across many sectors of society."

Many of us will have felt the pull of the sea when we need to restore or recoup. Just looking at water has a psychological restorative effect and getting immersed in the stuff is a gravity-free break from the rest of the world. The sound of water masks other noises, so that if we want to talk to someone (without shouting) in water, we have to be fairly close. Often, fairly naked, too. It's no coincidence so many coming-of-age novels or movies include a swimming pool or sea scene: water is sexy as hell.

Our bodies work differently in water. Many of the aches and pains of daily life disappear as we become swirling, twisting, writhing mermen and women in water, regardless of what we look like on land. We stop being our jobs, our responsibilities, our *thoughts*, even— surrendering to the elements and being just . . . us. Which is therapeutic.

It's also cold. At least, round my way. But this, too, can help. Several studies, including one published by the *British Medical Journal* in 2018, have found cold-water swimming an effective treatment for depression.[29,30,31] This is thought to be because immersion in cold water activates stress responses in our body, increasing our heart rate and blood pressure and releasing stress hormones. Repeated exposures start a process of "habituation" as we get used to it. Once we've dunked ourselves mercilessly into water below 59 degrees Fahrenheit a few times and "coped," we blunt our response to other daily stresses.

Two friends who've recently suffered from depression and bouts of ill health swear by cold-water swimming, insisting that the cortisol surge followed by feelings of euphoria once they're back in the relative warmth of "air" give them a high like no other. I join them one evening and peer anxiously into the inky sea. A quick check of the thermometer on the pier confirms my worst fears: 42 degrees Fahrenheit. There is a strong desire to run away and I consider how counterintuitive it is to even consider disrobing and jumping in. But thanks to peer pressure, I know: I'm doing it. Wearing nothing more substantial than a bikini (this is modest—the friend recovering from a mastectomy goes nude) we take the leap.

For a moment, there is panic. The cold bites at my skin. I flail about for a few seconds, struggling to gain breath/purchase/a sense of orientation, then scrabble for the slimy seaweed-strewn ladder at the end of the pier and get out, skin burning. It's not exactly "swimming," more "momentary plunging." But as soon as I am on dry land, I notice three things: first, that I am no longer cold, despite being near naked in an air temperature of only 46 degrees Fahrenheit. Second, my skin has turned the color of freshly sliced beet. *Attractive*. Third, I feel amazing. Despite the beet-colored skin and the undeniably "salty seadog" tang that has taken up residence in my nostrils and the back of my throat. The others emerge after a swim, bodies gleaming, slick with water. They are beaming. Buzzing. Loud, even, with exhilaration at what their bodies—and minds—have just accomplished.

I am persuaded to try again the following week and manage, according to skinny-dipper's watch (yeah, that's right: naked but for her trusty Apple Watch), TWO WHOLE MINUTES! I won't say that the experience is necessarily enjoyable (it isn't) but I'm glad that I've done it afterward. This is how I feel about most forms of exercise. I am yet to find anything that I actually enjoy—a problem since all the studies and experts agree that finding something we take pleasure in makes us more likely to stick at it.

Wrapped in a towel, flask of coffee in hand, I wave from the pier at the others, still swimming, and notice a whole world of recreation taking place on the water. From kayaks to rowing boats, windsurfers, sailors and even something I am reliably informed is called "stand-up paddleboarding," or SUP. It looks a lot like people are walking on water, Jesus-style, serenely dipping a stick into the sea at intervals. Then I notice that each "Jesus" is balanced on what looks like a surfboard. The whole thing looks *lovely*.

I am not a surfer. I am not cool or sporty. I was picked last for teams at school and didn't "do" exercise growing up (too many extracurricular activities on the calendar). My relationship with exercise in my twenties was one of punishment and a way to be less. Small, thin, unobtrusive. Running with balls has never appealed. But this? This!

Skinny-dipper knows a little about stand-up paddleboarding and so educates me on the act of propelling oneself on a floating platform with the help of a paddle. I learn that variations include sea surfing, white-water SUP, paddleboard yoga and even SUP fishing—all intended to promote physical health, serenity and a deeper connection with nature. There is, apparently, a club nearby where I can try it out with some very kind and patient people who will get me started.

"What are you waiting for?" she asks.

Nothing. I'm waiting for nothing. After being on bed rest, I swore I would move whenever I could. I've also been told that I'll probably need an operation to repair my stomach muscles after carrying twins, but a report from the American Council on Exercise[32] says that stand-up paddleboarding benefits abdominal muscles. So I try it, taking up the paddle in a last-ditch attempt to avoid the scalpel. I wobble at first. Fall in a lot (cue beet skin). Then, finally, I am up. And moving, under my own steam. At my own pace. I love it, in a way that I never thought I'd love any form of exercise (I always thought the people who said they did were in denial, mistaking "exercise" for "cake"). Getting out on a SUP is an opportunity for a mini adventure in beautiful surroundings whenever the wind is low enough. I barely work up a sweat and I'm far from ripped, but I do feel stronger.

Scientists have found that combining exercise with time spent out of doors may add extra benefits. A study from the University of Canberra found that getting out and getting active simultaneously could help reduce anxiety,[33] while researchers from the University of Exeter found that so-called green exercise decreased tension.[34] But it's not just that exercise makes us feel good: *not* exercising has been scientifically proven to make us feel bad. And can even lead to depression.

Dr. Brendon Stubbs is unique in his field for being a physiotherapist who has become a leading expert in mood and psychology. Stubbs was working on a psychiatric ward as a junior physiotherapist when he noticed that a lot of patients were sedentary, sitting still for most of the day. "It was the early 2000s but there was emerging evidence on the impact of a sedentary lifestyle and health," Stubbs tells me when I call to find out more. "So I just thought, What if I asked these sedentary patients to wear pedometers?" He started by looking at their average step count and then asked them to do 10 percent more. "So if they were doing 500 steps a day, I'd say, 'Add fifty.' We're talking small changes."

So categorically not the 10,000 steps we're all told we should be aiming for?

"No," says Stubbs, "that's an arbitrary number a company came up with in the run-up to the 1964 Summer Olympics in Tokyo, when they wanted people to buy more pedometers and so encouraged this high-level step count." This is a myth Stubbs is eager to bust, since although 10,000 steps a day is fantastic, for many of us it's not achievable. "There is no binary cut-off and there's a great deal of evidence that even low levels of activity are beneficial," he says.

Understood.

Stubbs noticed that increasing activity levels started to make a difference to patients' mood: "I could see that in practice, time and again, if you move, you feel better—and if you don't move, you start to feel worse."

And, er, I ask tentatively, aware it's been a while since I last moved, how long do you have to skip exercising for to start to feel worse?

"A week," is the answer.

Just a week?!

"Randomized control trials—the top-tier evidence-based trials here—show that after a week if you don't move, if you force someone to be sedentary, they will feel worse," says Stubbs. So if we're holding it together and don't have underlying health conditions, and we stop exercising for a while, we will start to feel bad. And the way society is set up, a sedentary lifestyle is alarmingly easy to access. In fact, it's pretty much the default. We've come a long way from our hunter-gatherer past, via manual labor in the industrial age, to a point where cars, computers and labor-saving devices have taken away much of the physicality of work. Now, it's fully feasible to stay "virtually connected" and never leave the comfort of our own couches. This is a problem for our mental health, says Stubbs. So next he flipped this thesis on its head to look at whether we can *stave off* depression and even manage low moods by keeping active and doing light exercise.

In 2016 he studied thirty randomized controlled trials that show how exercise could lessen depressive symptoms over a twelve-to-sixteen-week period.[35] "We found that exercise has been proven to be more effective than medication in some cases, and as effective as CBT," says Stubbs. This is big (huge) since getting active comes with none of the side effects associated with antidepressants and is easier to access than therapy. He adds: "I would say medication is really important to people and saves people's lives, as does CBT—but lifestyle interventions can, too."

Even when we're not suffering from depression, exercise can stop normal sadness and low moods tipping into something more serious. In 2018 Stubbs and his colleagues did a meta-analysis and found that exercise decreases the risk of developing depression, regardless of age and geographical region.[36] "We found that higher levels of physical activity were protective against future depression in children, adults and older adults," says Stubbs, "across every continent and after taking into account other important factors, such as body mass index, smoking and physical health conditions." In 2020 Stubbs and his colleagues published new research confirming that light activity is also good for preventing mental health problems in teenagers.[37] The evidence now exists, in black and white, that exercise works to both prevent depression *and* help treat it.

So why don't more of us know about this?

"I think it's fair to say that lifestyle interventions have a less powerful lobbying voice," says Stubbs. Whereas, "there is a very powerful body of pharmaceutical companies."

This is both disheartening—that we live in a cynical, money-driven world—and a source of hope. Hope, because if there's not a gap in evidence, just a gap in disseminating that evidence, it's something we can all play a part in.

I start tuning in to other people who've opened up about exercise helping with mental health issues. Celebrities like Demi Lovato, Olivia Munn and Khloe Kardashian are all helping to break the stigma. I'm heartened further when I see that *Girls* star Lena Dunham posted on Instagram:

Promised myself I would not let exercise be the first thing to go by the wayside when I got busy with Girls *Season 5 and here is why: it has helped me with my anxiety in ways I've never dreamed possible. To those struggling with anxiety, OCD, depression: I know it's mad annoying when people tell you to exercise, and it took me about 16 medicated years to listen. I'm glad I did. It ain't about the ass, it's about the brain.*[38]

Getting out and getting active now feels nonnegotiable for me. Being sad isn't a hall pass—it's not a get-out-of-jail-free card. We still have to do stuff. It takes work to stay "good sad." But not *too much* work.

"The most important thing to bear in mind is that we're not talking anything extreme," says Stubbs. "If you do 150 minutes of exercise a week, or 20 minutes a day, you have a 30 percent reduced risk of depression."

A reduction of a third is astonishing. I tell him I'm persuaded.

He tells me that we can do more exercise if we feel so inclined (Stubbs does): "But there's an upper ceiling of about 300 minutes per week," he says. No studies directly compare aerobic versus non-aerobic activity, "but I would say that they are broadly just as effective as each other," says Stubbs. So it doesn't matter what we do: "The most important thing is finding something that you enjoy—because then you'll stick to it."

I screw my courage to the sticking place and swear I will take action. I SUP whenever the weather allows and on other days, I walk—coming as it does with the added benefit of needing no kit and no preparation. I clock thousands of steps each day, taking new routes to go about my daily business, enjoying new sights and inhaling new scents. I pass a particularly pungent magnolia tree on the way home and become temporarily obsessed by it. I feel the sun on my face and find that walking makes me feel better. It helps me sleep more deeply at night and even helps me think more clearly during the day. Walking has a long association with creativity and Aristotle founded the Peripatetic school of philosophy essentially "walking and talking" back in the third century BC. The Danish philosopher Kierkegaard (him again) wrote in the nineteenth century: "I have walked myself into my best thoughts, and I know of no thought so burdensome that one cannot walk away from it."[39]

Apple CEO Steve Jobs was an eager walker and believed fresh

air and exercise helped him come up with his best ideas. A 2014 study from Stanford University confirmed that walking does indeed have a positive effect on creative thinking.[40] Just as sadness is the problem-solving emotion, it may be that walking is the problem-solving activity—one that's reassuringly undaunting when we're feeling low.

After six weeks of rain, thunderstorms and biblically bad weather (Oh hi, global warming) my handy weather app assures me that today there will be no wind. And better still, sun! I pull on neoprene boots and drive to the "surf club" (whatever mental image this conjures up, dial down the glamour x100). I unlock the rusty padlock to the shared shed where I keep my board, then set off. I am free. I am me.

An immense liquid mirror is pierced by a single stroke as I launch onto the sea's glassy surface. It's peaceful here: just me, my stand-up paddleboard and a seal that's taken to following me (I'm calling him "Derek"). The sun creates orange slices across the water as I follow the frilled coastline, breathe in the tangy sea air, and feel utterly . . . *okay*.

20

Get Even, Mind

Exercise makes a difference. Fresh air helps. But there's another thing, too. If we want to allow for normal "sad" and stop it tipping over into something else, we have to *get even*.

T once took a photograph of me in bed without my knowledge. We'd just moved in together, it was past midnight, I wasn't wearing much, and it wasn't nearly as creepy as it sounds. In the photograph, I am asleep ("still creepy," T points out. I can explain . . .) surrounded by sheets of heavily etched paper, laptop open about an inch from my drooling mouth, curled up in the fetal position. I look very much like a chalk outline from a crime scene, only with a pen in one hand and a BlackBerry in the other (yes, I know: this dates me).

I had been working too hard, staying late at the office, trying to start a new project in my own time, "failing" at fertility treatment and "failing" at seeing friends, T, or much of anything beyond the screens that surrounded me. As we've now established, I have a tendency to choose "work" over "life balance," so this photograph serves as a useful reminder of all that goes wrong when I succumb to this compulsion.

I was getting through the days on a combination of caffeine, sugar and shopping (bribery). I was getting through the nights courtesy of Sauvignon Blanc and sleeping pills. I was exercising: taking classes I didn't enjoy with words like "pump" in them at the gym underneath my office. I was reading, too. I once got mercilessly mocked for turning up at a *Sex and the City* premiere with a copy of the collected works of Anton Chekhov (to be fair, it was *Sex and the City 2*: I needed it). I was doing a few of the "right" things to avoid letting normal sadness tip into something else. But I wasn't looking after myself. Not really. Because there's more to being sad, well. The photograph of me + laptop + 1 a.m. drool reminds me that we need balance in our lives.

We need to get even.

"Even" has never been "cool." No one uses the phrase "even keel" to describe the exhilarating things in life. "Even" has a poor PR team. We don't fall "evenly" in love. Or have a dramatically "balanced" breakup. There aren't many pop songs written about "balance" and few Hollywood A-listers get caught out on a "balanced" bender, enjoying one drink responsibly before hitting the comforter hard. Because balance *isn't* exciting—but it is important.

Remember burn-out? Remember its nefarious symptoms of exhaustion, overwhelm, headaches and stomach pains? Some of these probably feel alarmingly familiar right now. Americans famously work some of the longest hours in the world and many admit to being "so busy" that even taking a break for lunch feels impossible, with the sad desk-lunch increasingly ubiquitous. But long hours don't deliver productivity, value or personal fulfillment; instead, they're a health hazard. Studies show that overwork leads to burn-out, poorer judgment, chronic fatigue and ultimately shorter lifespans,[1] says Alex Soojung-Kim Pang, visiting scholar at Stanford University and the founder of Silicon Valley consultancy Strategy and Rest (he was previously deputy editor of the *Encyclopaedia Britannica* so, you know, he *knows* things).

"The research on this has been very consistent, and it's literally been going on for more than a century," says Pang when I get in touch to find out more. He and I have spoken a few times over the years and I'm always interested in hearing more about his research. "Today," he says, "people are reaching the physical limits of their ability to work." As a result, a growing number of companies, often led by people who are veterans of Wall Street and Silicon Valley, are consciously trying to find new, more balanced ways of working.

Presenteeism is a big issue. "We don't measure productivity by how many acres we've harvested anymore, so the amount of time we spend working becomes a proxy," explains Pang. Performing busyness is a way of fitting in and when everyone does it, living a more balanced life looks like slacking. Overwork as a *choice* (as opposed to slaving away for subsistence wages) has been part of Western society since the Industrial Revolution. Some feared that automation would create the problem of "excess" leisure time. Needless to say, that didn't happen. Instead we all worked more—and then came the 1980s.

Thanks to computerization and globalization, managers could demand more of employees under the threat that jobs could be given to someone else. So the pressure piled on. And we took it, buckling under the strain, but shouldering the burden all the same. "Only this isn't sustainable," says Pang. We may think we can handle overwork in our twenties, but this, crucially, is because we're *in our twenties*.

"There is a steep professional learning curve early in our career that can be useful," accepts Pang, "and working like this is more sustainable when we're younger. But then there's an enormous body count of people who don't make it at all." I think of several friends who found the pressure of professional life unbearable in the first decade after leaving school or college. This is largely because the working conditions and hours they were expected to work were, indeed, near-impossible to bear.

"We all overestimate the degree to which we can handle a heavy workload," says Pang. "Especially as our lives change and especially as we become parents." In fact, he's observed that: "Society dangerously underestimates the importance and difficulty of parenthood." Pang, a father of two, says: "Children are the most important thing I've done in my life, but they are vampires. *Cute* vampires. But vampires nonetheless." I am always ridiculously thankful when someone else articulates the combined gratitude and daily despair and/or fatigue that is parenting. I tell him I was vomited on, twice, last night ("once, in my *ear*") and have five kids over for dinner later.

"It's *extremely* hard." Pang nods. "There is a study showing that women today take more time with their children than their grand-mothers did. That is . . ." I'm waiting for some sort of technical/Silicon Valley phrase but instead he says: "bonkers."

Yes, I nod: *That*.

"The idea of the 'constant intervention mode' of parenting to demonstrate that you're 'a good mother' is yet another trap for women," he says. He's right.

Our parents' generation had different expectations. "My dad grew up in Japanese-occupied Korea," Pang tells me, "my mother was dirt poor in West Virginia." When Pang was a child, his father was a college professor and spent a lot of time in Brazil, studying local records and land offices. "So we went, too," Pang says. "I grew up with the model of parenting that you could organize your life around your work—and the children could come along. Turning your kids loose and asking them to be home for dinner was about as good as it got." This is a helpful reminder, but it takes time, effort and ovaries of steel to fly in the face of modern society's expectations of parenthood and strike out for a better balance in our lives. Interestingly, Pang is now a "whatever gets you through the day" kind of parent.

"I'm much less conservative about screen time now, for example, and I don't see giving an iPad to a child on an airplane as the *worst* thing," he tells me, sharing that: "When I flew as a child, the stewardess would ask my mom if she wanted to give me a sleeping tablet. 'Drug them' was the standard then. So actually, watching *Mulan II* again? It's progress . . ." Today, he takes the view on parenting that, "If your children do not become child soldiers, serial killers or perish from communicable diseases, you're doing a good job."

Without wishing to instrumentalize my kids, I'm also conscious that being a parent makes me work less, and that this is probably a good thing. Okay, so I'm not using that free time to rest. But caregiving does give me a different priority. The trouble with defining myself in terms of my work and productivity in my former unenlightened (ha!) life was that every failure in my job was necessarily a failure in *me*. The failure *of* me, even. But we are not the sum of what we do. We are not worth any less because we are not achieving all the time. We have value, regardless of our output, bank balance, job title or role.

We all need something beyond ourselves to believe in—a purpose, something we're passionate about. Something to get us up in the morning that isn't our job and can stop us working too much. Because Pang's central message seems to be: if we *can* do less, we *should* do less. And rest.

"We need rest to make sense of our lives," says Pang. "Deliberate rest," as he calls it, is the true key to productivity and can give us more energy, creative ideas and a better balance in life. To ensure that he walks the walk as well as talks the talk on rest, I ask about Pang's schedule. He assures me that he regularly naps and refuses to over-schedule himself: "And it works."

I'd like to "rest" more. I'd love to "nap," but my Catholic guilt combined with Protestant work ethic (a heady combo) makes the idea seem ludicrously indulgent. Like Boxer in George Orwell's *Animal Farm*, my motto to live by has always been "work harder." I'm also conscious of the 10,000 hours theory—whereby it takes 10,000 hours of practice to get really good at something, as popularized in Malcolm Gladwell's book, *Outliers*.[2] The 10,000 hours rule is based on research by psychologist Anders Ericsson, now at Florida State University, who studied violinists at Berlin's Academy of Music and found that the best performers had all totaled 10,000 hours of practice each.[3] I know, logically, that there has to be some natural talent or aptitude. I can hit a ball as long as I like: I'm never going to be Serena Williams. But the gist remains: hard work = success. No pain, no #gains.

"So shouldn't we work hard? What about Malcolm Gladwell's theory?" I ask.

Pang tells me about a lesser reported part of Ericsson's study that strangely (or not strangely at all, depending on your perspective) contradicts the "work harder" thesis. In his study of violin students, Ericsson found that even though the most ambitious students needed 10,000 hours to become world-class performers and practiced more deliberately than their classmates, they also *slept* more.

"An hour per day more than average students," says Pang. "They took naps in the afternoon, and they were careful to get enough sleep at night."

Wha . . .? I make a sound like a distressed gull.

"It's true," says Pang, "they rested more."

This blows my actual mind. Once I've recovered, I ask Pang if he's surprised that the "rest" part didn't get picked up.

"'Yes' and 'no,'" is the answer. "There are two paragraphs on this in the original article. It's very easy to gloss over the sleep and the 'rest' part," he says. This is because "action" is prized in our culture whereas "inaction" is not. Or rather, what passes for inaction but is actually the time our neurons are restoring us to factory settings and our cells are rebuilding, recovering, readying us to fight another day. Subsequently, the nappers of this world have their snoozes airbrushed out of history. Pang references Roy Jenkins's biography of William Gladstone.[4] "He talks about him [Gladstone] working nonstop but then completely glosses over the fact that he goes hiking in Sicily for a month. Or that he reads *The Iliad* so closely, he notices that they never use the word 'blue.' I mean, really! You've got to be reading a book pretty deeply to realize that there's a lack of *blue* in it!" But, as is often the case, it was Gladstone's work that was prioritized and that is valued more highly in our culture. "Even really sympathetic biographies tend to focus on work," Pang has noticed.

But rest matters. Rest is when the good things happen, like restoring body and mind. Like friendship. Or love. Or great ideas. The fusion reactor model was conceived while skiing. The Beatles classic "Yesterday" came to Paul McCartney in a dream. Russian chemist Dmitri Mendeleev thought up the periodic table and the American inventor Elias Howe invented the sewing machine, both while sound asleep.

Pang now recommends we prioritize sleep as well as regular naps, where possible, and planned periods of total rest. He wants us all to commit to taking a week off every season to recoup and restore. "Scientists have found that the happiness and relaxation we feel on vacation peaks after about a week," says Pang, "while the psychological benefits of a vacation last up to two months. So the ideal vacation schedule is one week off every three months."

I knew dream doctor should have packed me off in a yurt some-where . . .

And in an ideal world? Pang says he'd prefer us to work just four days a week. "This has been proven to improve our personal life, by giving us more free time," he says.

"So, four days of work a week then the rest for rest and play." This is something Scandinavians are already considering, with Finland's Prime Minister Sanna Marin backing the idea. "But it's not just a

Nordic thing," says Pang, "it's happening all over the world. This is an important takeaway, since in the US, when we hear about something happening in the Nordic countries, it may as well be happening in Middle Earth. People think of the Nordics like these Hobbits, in a place where cool, magical stuff happens."

"Do they mention the taxes?" I ask.

"Not so much," he admits. But, he tells me, the four-day week is already happening in pockets internationally. "And it isn't confined to industries that already have a reputation for work-life balance." Microsoft Japan found that staff productivity leaped by 40 percent after a one-month trial of the four-day work week over summer 2019.[5] New Zealand-based finance company Perpetual Guardian attracted worldwide attention when it slashed the working week to four days— and saw a 20 percent rise in productivity, increased profits and an improvement to staff mental health.[6] They're working smarter, not harder. "For years there was this assumption that to succeed in modern life you had to sleep with your smartphone under the pillow—to be always on call, always working," says Pang, "but it's not so." In fact, we do better by avoiding our phone as much as possible. In his 2013 book, *The Distraction Addiction*,[7] Pang cites studies showing that a majority of workers have only three to fifteen minutes of uninterrupted working time in a day, thanks to smartphones. We also spend at least an hour a day—or five weeks a year—dealing with pinging, whizzing or vibrating distractions before getting back on track. According to a recent *Harvard Business Review* article,[8] a single text that takes on average 2.2 seconds to read can double our work error rates and it takes eleven minutes to regain the "flow state" achieved before the distraction. Needless to say, our productivity is at risk.

It's fine! You say: *I can multitask!*

You can't.

Research from Stanford University shows that multitaskers are markedly less productive than monotaskers and typically experience difficulties concentrating, recalling information and switching from one job to another.[9] Multitasking with screens only makes things worse. "Media multitasking" is associated with symptoms of depression and social anxiety,[10] a reduction in both short-term and long-term memory,[11] and could even be changing the structure of our brains. A 2014 study from the University of Sussex found that individuals who engaged in more media multitasking activity had smaller gray matter volumes in their anterior cingulate cortex[12]—the part of the brain that's linked to the regulation of mental and emotional activity.[13] Oh, and research from the University of Pittsburgh suggests that if we're on seven or more social media platforms, we are *three times* more likely to suffer from depression and anxiety.[14]

Thankfully for me, I can't even think of seven social media platforms (*seven?!*). But still, my love of the "likes" isn't massively healthy. Before the smartphone era, we spent mere minutes on our phones each day. Now, it's an average of three and a half hours.[15] Sometimes, I ignore my children to look at pictures of my children on my phone. Like a deranged person.

As we've discovered, the more time we spend online, the more likely we are to feel isolated, stressed and depressed. The digital consumer intelligence company Brandwatch published a report in 2019 chronicling the emotions of the British public as expressed online and found that peak "sadness" occurred at around 8 p.m.[16] This sounds about right. The day is done: work and/or family commitments have been taken care of. Eight p.m. is the time to sit and scroll through my phone to see what everyone else has been up to on social media. Eight p.m. is the time to discover that, lo and behold, "everyone else" appears to have been having the time of their lives. Eight p.m. is the time for that stomach-sinking feeling of "not good enough."

In his 1846 essay, "The Present Age," Kierkegaard analyzed the philosophical implications of a society dominated by mass media (sound familiar?) and wrote with disapproval about his contemporaries' curiosity with the trivial. He wrote about how such preoccupations made us all "lesser." Old Kierkegaard, it seems, had a pretty good handle on Facebook, 150 years before any of us had started poking each other (remember that? *That* was creepy). I ask Henrik Høgh-Olesen—the Kierkegaard expert—what the famous Dane would have made of the smartphone era and he hints that the nineteenth-century philosopher would have despaired at the way social media has taken over our lives.

"And speaking personally, as an evolutionary psychologist, there is a stronger basis for sadness now than ever before," he tells me.

How so?

"Well, in ordinary primal hordes, there would be between thirty and sixty people," he explains, adding, "So me and you would do okay—"

"Thanks very much."

"I mean, sure, there'd be some people prettier and some cleverer—"

All right, Henrik . . .

"—but we'd do okay. Now, though, our peer group is *everyone* we see in the media," he goes on. "Our peer group is the 1 percent of the planet who are beautiful and successful and geniuses—people who are so unlike most of the other people out there. And yet they're still setting the standard for most of us." If we strive for this, Høgh-Olesen believes, "we will be sad—no question."

If we don't get off our phones and get some balance in our lives around social media—if we don't get even—we will be sad. "No question." And studies also show that our reliance on smartphones and social media makes it more likely that normal sadness can tip over into something more serious, like depression or anxiety. So we should all want to make changes.

Pang recommends baby steps: "Put the phone screen down on the table while you get coffee. It's a little act of resistance, like saying, 'I'll get to you, when I choose.'" Experts also advise us to switch into airplane mode whenever possible, turn off notifications and ban smartphones from our bedrooms at night. The psychologist Philippa Perry goes even further, writing in *The Book You Wish Your Parents Had Read* that parents should keep smartphones out of sight when children are around: "We know alcoholics and drug addicts do not make the best parents because their priority is always the substance they're addicted to, so that children are denied a lot of attention they need. I'd say phone addicts are not so far behind."[17]

Since Pang first studied the impact of smartphones on our culture and our brains in 2013, more people have started taking smartphone overuse seriously. "People are aware on a personal level what they should be doing," he says, "they're just not *doing* it." He doesn't hold us entirely responsible here: "There are powerful economic incentives to resell our attention," because "people continue to get rich off it." Big companies pay for our eyeballs. The price of our attention is high because this way, organizations can know more about us and sell us more "stuff." My Instagram, Twitter and Facebook feeds are full of #ad and #gifted, no matter how hard I prune. Social media posts about "stuff" we can own seem in poor taste in these days of recession, unemployment and the fallout from a global pandemic (You too could have something shiny while others do not!). I know this. I feel it. And yet every time I go on Instagram, I leave several rabbit holes later with a vexing sense of dissatisfaction and the vague idea that if I just had boots like @hotmomblogger (I made this up, there may well be one called this) my life would be infinitely improved.

I read John Berger's *Ways of Seeing*[18] as a student and reflect that we have learned nothing in the past fifty (fifty!) years. Berger wrote of publicity and advertising that: "It proposes to each of us that we transform ourselves, or our lives, by buying something more. This more, it proposes, will make us in some way richer—even though we will be poorer by having spent our money."

The whole point of publicity and advertising is to encourage us to change our lives by buying something—to tell us that only when we've bought *the thing* will our lives be complete. That's not to say that we wouldn't enjoy the thing. Things can be great. We love things (especially T, who, while I am writing this chapter, has just ordered a new impractical chair for our house that I was not consulted on). Things are nice. But we don't *need* the things.

They may call it "retail therapy" but the "therapy" part is fleeting and phony. As Berger says: "The publicity image steals her love of herself as she is, and offers it back to her for the price of the product." If we allow advertisers to steal our sense of self-worth, our normal sadness can tip into something more serious. We have to look after ourselves—and none of us need so much "stuff" anyway.

The majority of us worldwide say that we could live contentedly without most of the things we own, according to a "prosumer" report by Havas Global Comms, one of the world's largest global communications groups. In 2016 even the furniture giant IKEA's head of sustainability declared: "We have probably hit peak stuff."[19] And when the Swedish flat-pack folk are telling us to take it easy and get even, it's probably time to get even. The Swedes tend to be fairly good at balance on the whole (tiny tea-lights purchased high on meatballs aside). The Swedish term *lagom* is all about living with balance and is used to describe a type of functional minimalism—an approach to life that most Swedes are taught from birth. One Swedish friend told me that his first memory from childhood was being asked, "Have you eaten enough?" and answering "*Lagom.*" A parent might ask a child, "Do these clothes fit you?" to which they might answer, "Yes, they are *lagom* big." *Lagom* means "sufficient" or "enough" and exemplifies the Swedish outlook on life and the realization that working more to earn more to buy more "stuff" is a fool's game (especially in Scandinavia with the taxes). The Norwegians have *passelig* or *passé*, meaning "fitting," "adequate" or "suitable"—so they might say the weather was "*passé varm*" (though this is unlikely in Norway) and a house can be "*passé stor*" or the "right" size: adequate. The Finns, my friend Marianne assures me, have *sopivasti*, which comes with similar connotations of "just right." But it's not only the Scandinavians who get the idea of "enough." In Thailand there's *por dee* or "it fits." One Thai teacher reports having *por dee* money to live comfortably with what she needs—a lot like Swedish *lagom*. Similarly, a dress can be *por dee* with a body and a job can be *por dee* with a lifestyle. Even the ancient Greeks had a version of this, with the sixth-century BC poet Cleobulus of Rhodes citing *metron ariston* or "moderation is best." Turns out that our glass may not be half full or half empty—it might just be filled *enough*.

Minimalism as a lifestyle choice has grown in popularity over the past decade but it's something that Joshua Becker began embracing in the 2000s. Becker thought that he was living the "American dream" with his wife and two children and a big house in Vermont. "Every time I got a pay rise, I bought a bigger house and the more we spent," he tells me. "We had a house full of stuff—not particularly fancy stuff, a lot of Target, but stuff, nonetheless. We weren't unusual," he points out, "but we had two kids—this was 'normal.'" Then one day he was cleaning out his garage and discovered that it was taking him rather a long time. All day, in fact, thanks to the sheer volume of "stuff" he had come to own. "My son was five years old at the time and he was asking me to come play with him in the backyard, as any five-year-old would," he says. "But I just kept saying, 'As soon as I'm done. Let me finish this and then we can play.'" Hours later, Becker was still working away at his pile of possessions. He looked at the "stuff" he'd spent all morning and afternoon tending to, and realized that it meant nothing to him. "Out of the corner of my eye, I saw my son, swinging alone in the backyard where he'd been all morning," Becker says, "and I just had this realization that everything I owned wasn't making me happy—worse, it was actually taking me away from the very thing that gave me purpose and fulfillment."

Deciding to make a change, he began to declutter, started giving away his possessions and set up a blog, BecomingMinimalist.com—a mission statement of sorts. He and his family downscaled to a house half the size. "Challenging at first," he admits, "but in a smaller space we were forced to learn how to coexist." As he began owning less, he found he had more time. "I also had more focus, less stress, less distraction—more *freedom*," he says. Then the recession hit in 2008 and the rest of the world started taking an interest. "People were losing jobs and houses and had less money and that economic downturn turned a lot of people into thinking about living a little more simply, whether by force or by design."

Today, the blog has 2 million followers; Becker has written five bestselling books on minimalism, and there are a slew of other lifestyle minimalists evangelizing for a better balance in our consumer lives, from the Minimalists, Joshua Fields Millburn and Ryan Nicodemus, to Leo Babauta, who somehow manages to stay minimalist with six children (six!). From Colin Wright, who only owns fifty-one things and travels all over the world, to *Stuffocation*'s James Wallman, who predicted that "experiences" rather than "stuff" are what really matter and has since been backed up by countless studies. Each of these people take different approaches, but minimalism has allowed them all to pursue purpose-driven lives that are better for their mental health, better for their finances and better for the planet.

According to the UN, we are living in "make or break" times for tackling climate change. To limit global warming to below 2.7 degrees Fahrenheit now requires a 15 percent cut in emissions annually and—

unsurprisingly—buying more "stuff" won't help.[20] Materialism is closely connected to mental well-being since depressed people buy more to make them feel better (see: shopping bribery). In what's become known as the "loop of loneliness," feeling sad makes us shop—and shopping makes us sad.[21] Researchers from Tilburg University in the Netherlands went one step further and found that valuing possessions as "happiness medicine" or as a measure of "success" increased loneliness the most.[22] Trying to sate or suppress sadness with "stuff" doesn't work. Whether it's material possessions or food. I know, I've tried (you probably have, too). Cake is nice. Cake's delicious, even. But it can't "fix" us. Turns out "cake as cure" comes with dubious medical efficacy, irrespective of what the internet memes say.

And now for a part of the book I didn't want to write . . .

21

Get Even, Body

Having worked for years to become comfortable with my body and develop a healthy relationship with food, this is the chapter I deferred writing for the longest time. I am still, in the present day, not immune to food issues. Just like depression, we're never "cured" with an eating disorder: we learn to manage it. Today, I eat pastries. I like pastries. Danish pastries are delicious. I want to be the kind of woman who eats pastries: I *eat* the pastries.

But I still don't like the feeling of flesh spilling over a waistband. Last week, T was interviewed by a fellow journalist and I heard myself commenting on how slim she was.

T looked at me sideways: "She spent five years in a concentration camp as a political prisoner."

"Oh. Right. Yes."

It wasn't my finest hour. But generally I do okay. I make no references to anything body- or looks-related around my children. I fill my plate at every mealtime and reliably clean it, too. Sometimes I under-eat. Sometimes I overeat. Always the sweet stuff. But I know enough by now to be wary of fad diets, trends or—most pernicious of all—absolutes. Orthorexia is very real and afflicted a good 60 percent of everyone I ever worked with on women's magazines. The body positivity movement has done a lot to counteract this. But what if we could eat well *and* feel better? Or at least, just feel "normal sad"? Because as unfashionable and as unwoke and as problematic as it is, it's true: we have to watch what we eat. Especially when we're sad and especially if we're prone to depression (essentially my Venn diagram).

"There is now a very large and consistent evidence base showing that the quality of our diets is very clearly linked to the risk for—or the presence of—common mental disorders," says Felice Jacka, professor of nutritional psychiatry at Deakin University in Australia, director of the Food & Mood Centre and author of *Brain Changer*.[1] A number of epidemiological studies[2] show that diet quality is directly linked to whether or not people suffer from clinical anxiety or depression—across countries, cultures and age groups[3]—and Jacka's now famous SMILES[4] trial showed diet to be an effective treatment for depression. I speak to Jacka one bright, crisp Wednesday after a week in which I haven't been eating well, and both my body and brain are letting me know that they're displeased with me. So I listen.

"We now have at least three randomized controlled trials on people with depression that show if you improve diet, it substantially improves depression," says Jacka, "and that seems to be independent of all of these other factors that we always take into account—like people's socioeconomic status, their education levels, health, how physically active they are, their body weight, those sorts of things." She's also shown that dietary interventions improve depressive symptoms in the *nondepressed*.[5] So while a diagnosis of clinical depression is at least five of the nine key symptoms laid out in the *DSM*, depressive *symptoms* could be just some of these—expressions of low mood that most of us will have experienced. In all scenarios, diet helps, and more so if we're women. Studies showed even greater benefits from dietary interventions for women with symptoms of both depression and anxiety.[6]

Diets heavy in whole foods work best for most of us and in one study, people eating the traditional Mediterranean diet, rich in vegetables, seafood and unsaturated fats, low in refined sugar, were half as likely to be diagnosed with depression over a four-year period.[7] But traditional cuisines of all varieties are preferable to a modern, Western diet of ultra-processed foods that now account for more than half of all calories consumed in the US.[8]

Sugar is a major problem with many of us consuming more than three times the World Health Organization's daily recommendation[9,10] and high-sugar diets are proven to lead to an increase in the same inflammation markers that are raised in people with depression. "Trans fats have also been clearly linked to depression," says Jacka, "now banned in many countries due to their associated health risks, but you can still find them around." Terms to watch out for are "hydrogenated oil," "partially hydrogenated" or "vegetable oil" on an ingredients list (olive oil is technically a fruit oil so a-okay).[11]

A straw poll of my social circle shows that a) most of us are woefully ignorant here and b) many right-on, well-educated friends are adamant that any attempt to "control" or modify their diet is another form of body shaming. But Jacka is unrelentingly body positive: "It's not about weight," she says on more than one occasion, "it's about diet quality."

"If you make the conversation about obesity," says Jacka, "then people are focused on the *wrong thing*—and they focus on something that's actually pretty hard to change. There's a genetic component to body weight that we can't do anything about. And there have been large-scale changes to our food environment, which are very difficult for individuals to fight against," she says. What is within our control, she says, is what we eat, *now*.

And food is a helpful way to keep balanced?

"Absolutely," is her response. "There might be some people for whom food is extremely important and is enough, in and of itself, to prevent—or cause—depression," she says, "but the way I think about it is more that if you have a healthy diet, it's like your foundation—your 'step zero.' And that means all the other risk factors, whether they're genetic, whether they're things that happen in your life that are really traumatic, things that are outside of your control—you're just going to be more resilient. You'll be more resilient to developing depression or, if you do develop it, it'll be easier to overcome."

She leaves me with what is possibly the best news I've heard all week: that hummus counts toward our daily vegetable intake ("It's true! Chickpeas are the best!").[12]

We can be sad—we can *feel*—but we may be able to avoid tipping over into something more serious by looking after ourselves better. By eating more hummus and less sugar (at least, this is my takeaway). Scientists at Princeton University found that the chemicals released by sugar consumption induce the same brain activity as heroin[13] and after speaking to Jacka, I become convinced of the efficacy of the very simple advice for diet and mood: go Mediterranean. So I do. After which, the night sweats I had assumed were the start of the perimenopause (really) miraculously disappear. I have a weekend where I go big on sugar and caffeine and don't sleep enough and drink more wine than I can handle and the next day I have the first panic attack in years. I forget my own pin number, get stranded at a train station far from home until a kind passer-by named Ken calms me down (thanks, Ken).

I get even again. When I eat a mostly Mediterranean diet, I am okay. When I don't, I'm not. I still eat whatever I'm given at someone else's home. I still go out, eat, drink and try to be social, wary of going against the herd. But if I don't regain some semblance of balance and get even pretty soon after, I'm in for a downward spiral where normal sadness can tip into low moods, or worse.

I spend a long time railing against this: feeling it's all too silly and superficial and why can't I burn the midnight oil, knock back espresso martinis and eat what I like and still feel well the next day, like the people around me? But I am not the people around me. I'm me. And this is how it is. I am forty and this is still a tricky thing to admit to. So I talk to someone who learned to go against the herd and come to terms with "balance" while still a teenager.

I first meet Ella Mills, founder of Deliciously Ella, at the start of 2019. At first glance, it's easy to assume that the tremendously success-ful, beautiful, wise-beyond-her-years Mills might not know much about sadness. But success is no fortification against unhappiness—and Mills has done "balance" even when it meant being at odds with what the rest of her peers were doing. She had no interest in healthy living growing up. "It never crossed my mind," she tells me. "I was a sugar addict—I'd eat gummy bears for breakfast." She grew up in a busy family as one of four siblings and left home to study art history. "I was having the best time," she tells me, "dating the cool guy, going out every night—all I cared about was what was happening tonight, who was going where." In the summer after her second year, she went to Paris and started modeling. Then things began to go wrong.

"Out of nowhere I suddenly became really ill," she says. "I couldn't get out of bed, and it took us about four months of all different hos-pitals, doctors, every test under the sign, including one ten-day stay in the hospital, to find out what was wrong with me." Mills was finally diagnosed with postural tachycardia syndrome or POTS—the same chronic illness that Hannah Lucas, founder of the notOK App has. She found it hard to tell people about what was happening, saying: "It was such a difficult issue to articulate that I didn't really want to talk about it with people." So she started to shut herself away and describes sinking "deeper and deeper" until, before long, she was suffering from "very bad depression" that left her isolated for years.

In desperation, she searched online for anything else that could help and came across the work of Kris Carr, the *New York Times* bestselling author who advocates a vegan diet. Mills tried it and began to feel better. She started a blog of her experiments with food and learning to cook, the now famous Deliciously Ella. "I explained that I wanted to learn to 'like' healthy food, 'because I'm sick and I want to see if it helps.'" Other people began to share their stories, too, "and you start to realize that everybody goes through real challenges in their lives," says Mills. This shift in her outlook helped her to cope: "The quicker you accept that life 'isn't fair,' the happier you're going to be," says Mills, "and I think I was struggling with that. But I do think it's a 360-degree approach," she adds, "because you can eat all the kale in the world, but if you're really struggling with your mental well-being, it's no good. You have to look after yourself like a whole person. You have to connect to how you actually feel and accept and deal with it."

The blog attracted 180 million visitors and in 2015 Mills published *Deliciously Ella: Awesome Ingredients, Incredible Food that You and Your Body Will Love*, which became a runaway bestseller. But Mills found the newly ramped-up media attention overwhelming. "I was feeling stronger," she says, "then people started commenting . . . on *everything*. It never bothers me if you have something constructively critical to say about something I do," she says, "but people were also saying things like, 'I hate your voice' or 'You're really ugly' or 'You're a spoiled brat.'" She was twenty-three years old and had just started dating her now husband and business partner, Matt Mills, son of the politician Dame Tessa Jowell. They got engaged and planned to open a chain of delis. Life should have been exciting, but instead she experienced crippling anxiety. "It was the sense of being watched, the sudden propelling into the public eye where everyone had an opinion of you—and the intense sense of vulnerability that came with it—that I just wasn't ready for or mature enough to handle," she says, adding, "it was one of the worst I've ever felt."

Then her parents told her that they were getting divorced. "We found out my dad had had a long relationship with someone else and was gay," she says. "That was a lot to digest." Then her business hit a bump in the road. "All start-ups hit different challenges at different points; it's not easy starting something from the ground up and we had a few real challenges," she says. Just when it seemed that the business was on track, with two delis, a collection of recipe books, a range of plant-based food products and a growing social media community, tragedy struck. "Matt's mom had these big seizures and was rushed to hospital. And we then found out later that she had a terminal brain cancer. She was with us for almost exactly twelve months from that day." This loss put everything else into perspective. Mills wrote of her late mother-in-law: "I've never seen love like I have since becoming part of this family."

"She was a total inspiration," says Mills, explaining how Jowell still managed to enjoy life until the end. "I remember we had one gorgeous summer afternoon, and Tessa said, 'What a perfect day.' And it was so interesting because she really meant it." She was dying, but she meant it. "It was such a good lesson of life: that everyone's together— that is a really, really good day. Especially if you know you don't have as many of them [left] as you'd like. Then why are you not appreciating it?"

This is a lesson in balance, too. Not denying our feelings or the sadness of life, but enjoying what is good at the same time.

"I do sometimes think 'getting on with it' is not the worst thing," agrees Mills. "There's a balance between the two that I think it's very important that we strike—because openness is very important and openness normalizes things. It is so normal to feel anxious. It is so normal to feel sad. It is so normal to feel not good enough. These are really normal human emotions. And I think the sooner we realize that that's what *everyone* feels, no matter who they are and what they do, the better.

"You're not just sweeping your feelings under the carpet, pretending they're not there, you're acknowledging how you feel. You are acknowledging the challenge. But at the same time, you're not going to allow it to define you." She has now produced five books, thirty plant-based products, an app, a podcast—and two daughters, Skye, born in 2019, and May, born in 2020. "There will always be sadness, but you're going to go on with it."

Psychologists describe this as "active acceptance"—a sort of comfortable compromise that goes against the grain of much of what we've been taught in the West. It's an acknowledgment that life will throw us curveballs but that we'll keep going, trying our best with whatever means are in our control, looking after ourselves and each other (more in Chapter 22). We can struggle for years to "fix ourselves" or "have it all" and feel frustrated by how much more we want to accomplish. Irrespective of how hard we work or how much time we spend with our family or friends, we may never feel as though we're doing "enough." So we have to find another way: to accept ourselves and what we have, *now*.

To address his own feelings of "enough," Harvard University's Dr. Tal Ben-Shahar identified five areas in his life where it was important for him to thrive: as a parent, as a partner, professionally, as a friend and in terms of his health. Then he looked for role models in each of these five areas—the people he knew who were nailing it. "While I found people who were doing some of these things well, none of them was a role model in all five areas—or even in a majority of them," he describes in *The Pursuit of Perfect*.[14]

This is a game changer. Try it and see. Think of an element of someone's life that you aspire to—envy, even. Then see if they're nailing it in the other areas of life that are most important to you. Chances are, they're not. Not even close. My top five are near identical to Ben-Shahar's—and no one I know is "winning" at parenting, work, relationships, health and friendship. It's also a useful exercise in reframing, maintaining some perspective, managing expectations, finding balance and appreciating what we've got. Because we could all do with finding our own "comfortable compromise."

Ben-Shahar drew up a bespoke game plan for his life based on what he discovered. After asking himself what would be the minimum he would be content with in each of the key areas of his life, he figured out that he could restrict his work to nine to five; manage three runs and two yoga sessions a week; a date night with his wife; and a night out with friends. This left five nights a week with his family. This, he concedes, was far short of his perfectionist ideal. But it was good enough. It was *lagom*.

I try to do the same, within the confines of the time I have available (and the fact that more than half the inhabitants of my house still need help wiping their butt). I can't cram in quite as much as Ben-Shahar, but I cover the basics that I need to stay sane and find the principle works.

We all need balance to be sad, well. We all need to get even, when we're feeling low. And no, it may not be exciting, but it can stop us falling into an enormous pit of despair. And that's a pretty worthwhile goal for today, for tomorrow, and for the day after that [*raises glass of water, toasts*].

22

Do Something for Someone Else

I F YOU'RE SAD and you just "do you," chances are you'll still be sad. And not the "good" kind, either: the niggling, panicky kind— riven with doubt and indecision and that feeling of "isn't there more to life than this?" If you're asking this question already, then the answer is: "Yes" and "Don't worry." Swiftly followed by: "Take my hand and let's do this together." Because in order to be sad, well, we need to pull back the camera: we need to do something for someone else.

If the past couple of years have taught us anything, it's that doing nothing isn't an option. Once we've rested, restored and fortified our- selves for civic life, we have to step up. Whether it's by taking care of our neighbors, or donating, or protesting, or all of the aforementioned and more. We are all activists now—or at least, we should be.

One of the resounding messages from the Black Lives Matter move- ment is that as a white ally, I/we have to "do more." Just being anti- racist isn't enough. I didn't choose to be white, but I can choose what I do now. I've been forced to re-examine my experience of being white and growing up in a culture where many doors were open to me only because of this. I acknowledge the systems and structures that I've benefited from and have woken up to the fact that it's not just up to the people who encounter racism to "deal with it": it's on me.

I'm trying to educate myself and work out the best way forward. My first instinct is to ask for more help from the Black people around me, but I soon realize that farming out the emotional labor for what to do is part of the problem. "We are tired," writes Nompumelelo Mungi Ngomane in a powerful piece for Medium entitled: "Dear White People: Now You Know and You Can't Pretend You Don't."[1] "We're still showing up for our white friends who want to know how they can help," she says, but "it's heavy and it's exhausting." I get this—I have to take it back on myself. Ngomane does offer some recommendations that she's happy for me to share and amplify, like making a recurring donation to organizations like Black Lives Matter (Global Network—blacklivesmatter.com), Equal Justice Initiative, MoveOn, Color of Change, Black Futures Lab, the Bail Project and the American Civil Liberties Union (ACLU). She also advocates: "Protest alongside people of color and, when you can, put your body between them and the police; boycott those beauty and fashion brands that are paying lip service to the [Black Lives Matter] movement until they pull up for all their customers; if you can vote, vote," and "when your Black friends do want to share their experience or correct you, listen." I try. I'll keep trying. We need to live by the principles of *ubuntu*. We have to "show up" for other people. We're meant to "show up"—it's hardwired into us.

"If you want others to be happy, practice compassion. If you want to be happy, practice compassion," says the Dalai Lama and altruism has been proven time and again to be good for us.[2] Studies show that doing volunteer work makes us feel better[3] and helping others improves our support networks and helps us to be more active.[4] Giving our time to do something for someone else, counterintuitively, makes us feel as though we have *more* time. In a study from the University of Pennsylvania, Yale and Harvard, researchers compared spending time on other people to spending time on ourselves, a windfall of free time, and even wasting time. They found that spending time on others increased feelings of time affluence, hands down.[5]

Spending money on other people has a similar effect—we don't feel poorer for having given some of our cash away, we feel richer. Harvard researchers found that "prosocial spending" makes us feel good and more bountiful, wherever we are in the world and regardless of our income or socioeconomic status.[6] Donating to charity makes us feel good, too, with a 2010 Harvard Business School study equating the rise in well-being achieved by donating to charity with a doubling in household income.[7] There's also a positive feedback loop, meaning we're more likely to spend on others in the future.[8] So prosocial spending increases well-being, encourages *more* prosocial spending, which in turn increases well-being, and so on.

The economist James Andreoni has developed a whole economic theory known as "warm-glow giving" to describe the emotional reward of giving to others.[9] MRI scans show that our brains literally light up—glowing with the pleasure of giving—and the inherent pleasure derived from "doing good" is also known as "helper's high." Yet surprisingly few of us are doing it. Volunteering levels have declined in the US, with fewer Americans volunteering and giving than at any time in the last two decades.[10]

Growing up, my Catholic education was crammed with stories of "doing good," from the celebrated Samaritan to "loving thy neighbor" or Mary Magdalene washing Jesus's feet with her hair (this one was more niche). Of course, the Bible also has numerous examples of conditional giving: doing something nice for someone because there'll be "payback" in heaven[11] or because "God'll get mad if we don't."[12] But there was also a sense that individuals should "do good," just *because*.

In my youth, I volunteered for Help the Aged, I served Christmas dinners in nursing homes.

Only in popular culture over the past couple of decades, the idea of doing good was replaced by the pejorative concept of a "do-gooder." A "do-gooder" was framed as a figure of derision (think Ned Flanders in *The Simpsons*). As we became more individualistic, no one wanted to be a "do-gooder"—or, worse, a "busybody," meddling or appearing overly interested in other people's business. The idea of doing things for other people as a blanket societal duty fell out of fashion. But this is a grave oversight because we *should* be more interested in each other's business. "Meddle" in it, even.

The religious historian and former nun Karen Armstrong campaigns for a return to the so-called Golden Rule: the idea that we should be active participants in our society and treat others as we'd like to be treated. In her 2009 TED Talk[13] she refers to the centrality of compassion in all the major world faiths, each of which has its own version of the Golden Rule. First propounded by Confucius five centuries before Jesus, the Golden Rule is the "source of all morality," says Armstrong. She urges all of us to get out there and do something for someone else—for kindness and empathy for its own sake. This has been the motivation for many of the people I've spoken to for this book who have moved on with (not "*from*") their own sadness to try to help others.

Richard Clothier, the spokesman for male infertility, now works to encourage other men to open up and helps run a support group for those experiencing infertility.

Since her son Willem died, Marina Fogle has become a vocal supporter of Tommy's, a charity that funds research into miscarriage, stillbirth and premature birth. She's also become a patron of a charity that supports families and educates professionals when a child dies or is facing bereavement. "Do I wish Willem was alive? Of course," she says. "But would I give back all that I have learned? All the strength the experience has given to me and all the amazing organizations we're able to work with and help now? I wouldn't want to go back and lose this now," she tells me, adding: "It feels useful to be able to do something."

We're well positioned to help other people when we're feeling sad as we're more empathetic and clear-eyed than we might be in happier times. The fundamental attribution error is the tendency many of us have to believe others are intentional when they make mistakes or say something wrong.[14] When we're sad, we're less likely to think the worst of people. We're also less susceptible to being dazzled by the halo effect, a cognitive bias whereby we think that certain people— usually the attractive[15] or successful—can do no wrong. Sadness gives us a less biased view of people and we seem to instinctively understand and appreciate that everyone is fallible. So sadness helps us to assess situations and the people around us more accurately and promotes a more attentive thinking style. We're better attuned to helping others when we're sad.

Involuntary founders of the Dead Dad Club, Jack Baxter and Ben May, are a prime example of this. Along with another friend, May and Baxter founded the New Normal,[16] a charity aiming to break down the taboo of grief and show that "it's okay to feel." Described as "a unique alternative to therapy," the New Normal seeks to connect young adults in need of support or experiencing difficult times. A lot like the Buddy System, "it doesn't matter who you are, if you experience grief you need help," says May. "We don't have all the answers. But we do this because it's a thing we should do. Because it might help other people."

We should help other people because it's the *right* thing to do and we have—sounds "old-fashioned word" bell—*morals*. All of us have convictions of what is good and most important what is *kind*.

Joshua Becker was faced with a paradox soon after he started

BecomingMinimalist.com. He wrote a book about his experiences that sparked a bidding war between nine publishers. "So we knew we were going to be paid more than we ever expected," he tells me. Becker is too delicate to talk figures but since the resultant book, *The More of Less*,[17] went on to be a bestseller, it's fair to say the sum was sizable. "It seemed to me the ultimate hypocrisy to use the book advance money—and do what? Buy more *stuff*? When it's a book about having less? It didn't feel right." So he put the advance and royalties into setting up the Hope Effect—a philanthropic charity to "change how the world cares for orphans."

As a newborn, Becker's wife had been left by her birth mother at a hospital. She was adopted almost immediately and raised in a loving family, but Becker knew that not all orphaned children had the same experience. "In fact, less than 1 percent of orphans are ever adopted," he says. "We've known for decades that it's pretty harmful the way most orphanages run—that neglect is harmful for brain development. So with the Hope Effect we wanted to provide family-based opportunities for orphaned children." Instead of large institutional group homes, the Hope Effect builds smaller homes, each made up of two "parents" and six to eight children. Large "families" by modern standards, but families nonetheless. "This way, children get more individual care and attention, along with stability and security," says Becker. At present, the Hope Effect is involved in providing nearly 100 homes for orphaned children in Honduras and Mexico, a couple of hours away from where Becker now lives. While researching orphan care around the world, he shares a story with me about an orphanage in Mexico where the children themselves are encouraged to do something for other people. I'm just starting to worry that this is all a little Oliver Twist when he clarifies, "And they do this because these children grow up in the worst circumstances—they get used to seeing themselves as victims. But then they start giving to other people. And they see they have something to give. By serving others they realize they can over-come that victim mindset. And so they feel better about themselves." Now, Becker says, this is a central tenet to his life approach: "Everything I do now is really aligned with that worldview." If we want to feel better, we have to do something for someone else. We pay our kindness and acts of service forward—and they really are contagious.

Researchers from the University of California, San Diego and Harvard found that cooperative behavior spreads from person to person. When someone does something for us, we are more likely to help others, creating a "cascade of cooperation."[18] So our acts of kindness spread to people we don't know or have never met.[19] As Aesop—of *Fables* fame—said: "No act of kindness, no matter how small, is ever wasted."

Our levels of kindness aren't set in stone and research shows that compassion—like patience—can be cultivated with training.[20] We can all learn to be kinder—we just have to start. Somewhere.

Speaking to friends, family and colleagues, from all over, I notice that those who appear most "together," with their emotions and well-being in a healthy state, all do something for someone else on a regular basis. One volunteers at a local asylum center. Another puts her crafting skills to good use and crochets the very same octopuses that helped soothe and keep my kids safe in the special care baby unit in the hospital (the twins ended up there, too). Since the inception of the Octo Project, there are now volunteer crafters in Sweden, Norway, Iceland, the Faroe Islands, Germany, Belgium, the Netherlands, Luxembourg, France, Italy, Turkey, Croatia, Israel, the Palestinian Territories, Australia, the UK and the US. I can't crochet for shit, but anyone who fancies it can follow the approved pattern in the Notes at the end of this book and donate the results to a neonatal ward.[21]

Medic turned comedian Adam Kay fundraises for the Lullaby Trust, the "very special" charity trying to reduce the number of SIDS deaths and support bereaved families. Kay held bucket collections for the charity throughout his tour for his book *This Is Going to Hurt*, raising almost $140,000. "They do such an amazing job for people at their very lowest ebb and I'm very proud to have played a part in the fundraising they do," he's said, adding: "I know what a difference it makes to the charity in terms of the helpline and support they give to families, as well as the research they do."

The historian Thomas Dixon is trying to help children learn about their emotions. He's working on a "living with feelings" project in schools, creating a series of lessons that he hopes will be "healthy and helpful" and encourage the next generation to come to grips with their "sad" a little earlier on. And we can all help to normalize the discussion of our emotions—good and bad. We can encourage the expression of feelings and be honest about our own. And we can all do something for someone else, not because we're getting something out of it or even because it feels good (though it does)—but because it's the right thing to do.

"I would argue for the intrinsic value of the benevolent act, irrespective of any emotional impact on the person responsible for it," explains psychologist and philosopher Svend Brinkmann, who believes: "You should do good deeds, not because they make you feel good but because *they* are good. Today we often tend to think, What's in it for me? or we do a cost–benefit analysis—we are rational egoists who do something non-egotistical only if it serves us," he says, "but that's a choice. We still *know* what is good."

I want to believe him (really I do) but . . . do we? Always?

"Yes," is the answer. "I tend to direct people to little, everyday situations that I believe they know of, where they can see the intrinsic value in doing good. Like spending time with your children."

I point out here that many people do this because it brings *them* joy. "That, and they worry that the children will be psychologically scarred if they don't."

He concedes that people may have mixed motives when it comes to their children.

"Okay then," he says, "helping others in need like being a good Samaritan, regardless of whether the Samaritan gains something. It's cynical to say that you would only rescue someone who was drowning because you wanted to 'look good.' Most people would say, 'Yeah, I guess that's true.' Just as two plus two equals four, you wouldn't think, I just want people to think I'm nice. No: you'd just rescue them. There are ethical norms."

When he explains this, he says, most people understand and come around to his way of thinking. "It doesn't take away from the fact that morally good actions make us happy—that's where we need to be quite clear. They *can* make us happy," says Brinkmann, "but that should not be the *motivation*."

I think about this. About the people I know who serve others and "do good," even when it might not be advantageous to them. My mother-in-law, for example. Until she retired, she was a physiotherapist for children with disability. Sometimes, the children had such severe disabilities that they died before reaching adulthood and my mother-in-law would always go to their funerals and visit the parents at home. She didn't have to do this: it wasn't part of her job. It certainly wasn't "fun." But she did it because it was the kind and compassionate thing to do. The *right* thing to do.

"Stoicism is grounded in duty," Brinkmann reflects. "The Stoics believed you should be a citizen of the world." But there's a realm in which he departs from their teaching. "Stoics give too much control to the individual," he thinks. "They say, 'You can't control the outer world, so control the inner world.'" This is a philosophy I have some sympathy with, I tell him. "But it disregards the fact that our lives are much more influenced by society and relationships," says Brinkmann. "Instead of building this in a citadel, we should build strong communities. We need to start socializing stoicism." There speaks a Dane, I can't help thinking. But even Denmark, he feels, could go further toward making kindness and good deeds a collective endeavor. "In Denmark we talk about the individual having to be 'robust' and to 'develop resilience,'" says Brinkmann, "but it would be better to build robust and resilient societies. To stand together." This is an admirable goal and one that—despite Brinkmann's critique of his homeland—I think Denmark does better than most. The sky-high taxes fund a welfare state that looks after everyone (at least that's the theory). We pay to care for people more vulnerable than ourselves—not because we know them or because it gives us a warm fuzzy feeling, but because it's the right thing to do. The good thing. Because we have a duty of care. As Meik Wiking puts it: "I believe we have a moral obligation to focus our attention to where well-being is most scarce and bridge the gap," since "well-being inequality has a larger negative impact on how we feel about our lives than does economic inequality."

Also: there's a lot that needs doing. Sticking our hands in our pockets and whistling is not an option. We have to care about other people and want to help them. That's the deal. We don't have to throw bags of flour out of the back of a van somewhere or join the Peace Corps to make a difference (though we can). But we can all offer micro acts of kindness, of service. When COVID-19 locked down much of the world, putting the vulnerable at even greater risk, we started shopping for them. Checking in on them. Buying them bread and milk when they couldn't get out of the house. We're not talking superhuman feats—we're talking being a nice person. We collectively achieved a fresh perspective and an understanding of what was really important to us. And we need to keep hold of this. The trauma of this pandemic will last for some time.

If someone ever says, "I'm scared," we help. That's part of what we do as human beings. When people are angry, we work out *why* they're angry and try to help them. We put a Band-Aid on the wound—not lemon juice. If someone is grieving, we let them. We say, "I'm sorry," and acknowledge it. We don't have to "fix it."

We have all experienced a global interruption and the weight of the world seems heavier than ever right now. But as well as the sadness, there has been a moment of bonding—a group collective witness. And it's a time for compassion.

After ten years of sponsoring other people's marathons, donating to charities and listing the Lullaby Trust as the main beneficiary of my will (so far, my assets total a well-thumbed bookshelf and a pretty jazzy pair of high tops, but I live in hope . . .), I want to do more. Speaking to Adam Kay while writing this book, I've realized that I can—and should—do more. And in lieu of many useful talents to bequeath them, I came to understand that there is one realm in which I may be able to help. So now I have trained to be a sibling supporter, available to help other siblings who have experienced SIDS in their family and are lost, bereft and struggling to know how to feel about it. "There is often some guilt around this," says Jenny Ward from the Lullaby Trust, "some stigma, this idea that it's not somehow legitimate to grieve as a sibling. But we know now that its effects are lifelong. It will be with you as a family on an ongoing basis." And she's right. "You might think, What would they have been like at this age? Would they have got married? Would they have had children? What would it be like to have an adult sibling with me now?"

I ponder all of these and more. But now, at least, I hope, I have allowed for some of these thoughts to take root and unfurl. I've felt, by now, all the feels—and I've worked on them. too. "This is important as we don't want someone to be triggered when they hear someone else's story," says Ward. "We need to wait until someone is ready."

I want to do something for someone else. I want to help. And I'm ready.

Epilogue

I'M WALKING TO the station in a "cool" part of London that looks a lot like most other parts of London except every outlet has the word "*dirty*" in front of it, the chain restaurants are painted black, and for some inexplicable reason there's a mural of two seagulls decapitating a doll. I have a feeling this is supposed to be "edgy" but it just reminds me of my kids' "robust play" and I miss them.

Time away from my family feels like a great idea on days when I want to run off and camp in a tent with no phone signal forevermore, but now that I'm here, all I want is to be with them. To assuage the guilt, I FaceTime and, unusually, someone locates the iPad in time from under a pile of something and answers.

T positions me on the kitchen table so that I can see one twin dressed as a lion, trickling milk out of his mouth while the other, smeared in what I can only *hope* is mud, eats an egg with her bare hands. I left in the dark, before the family was up. I've only been gone for five hours but already the house looks like we've been burgled and the general atmosphere is *Lord of the Flies* meets a Dali painting.

"Still missing us?" T asks, covered in distinctly amateur "clown make-up."

I nod.

The five-year-old, wearing only his pants, insists on speaking to me "privately," telling me, "I want to show you a secret in my bedroom."

O-kay . . .

The sun catches his auburn hair as he seizes the iPad and bounds up the stairs, his magnificent autumnal tones gleaming like a giant toffee.

He looks around and then whispers conspiratorially: "See that gap between my bed and the wall? D'you know what I keep there?"

I do not.

"*Boogers!*"

Oh!

"Old ones I don't need anymore! Ever since you told me not to wipe them on the wall by my bed"—It's true: I did make this request—

"I started putting them down there instead!"

Great! *That'll be a nice job for Daddy later* . . .

A pigeon defecates on me (lucky?), much to my son's hilarity, so I search for something to wipe it off with, say my goodbyes and hang up. Soiled but filled with love, the sun makes me blink back tears.

Every stage of life poses its own challenges. New babies are hard. Big "babies" are hard, too. Bigger children, I'm assured, are harder still, bringing new, bigger problems. Adults are hard. Navigating the turgid waters of parenting or marriage or human relationships, even, can be challenging. It's all *hard*: but we do it anyway, even when it makes us sad. There is meaning in pain and sadness. If we're sad or scared, it's a sign that we care: that we're connected. We need to experience all our emotions and live with our suffering—enduring rather than denying or anesthetizing it. We need to shake off the shackles of shame around feeling sad and allow ourselves to just feel it. To sit with our own discomfort. To flex that muscle and build up our resilience to "sadness." Because heavy lifting makes our muscles stronger. Uncomfortable? *Fine*. Awkward? Get used to awkward. Awkward's okay. Really.

Teaching ourselves to withstand discomforts in a low-stakes environment can prepare us for the whoppers that are wont to leap out at us, helping us to brace ourselves for the kind of sadness that comes like a blow to the back of the knees. We will all be bushwhacked by heartbreak, or pockmarked by the pain of everyday life not having turned out as we'd hoped. Life is a series of losses and love may be the greatest risk of all. None of us are immune from pain and we're fools to think that any amount of success or money or Instagram followers can ever solve our problems for us. We all have origin damage and—spoiler—it doesn't end well for any of us. As T likes to remind me in his "dour" moments: "None of us are getting out of this alive." And that's okay.

Sadness has a point. It's normal. And it can tell us when something is wrong, if we let it. If we're obsessed with the pursuit of happiness to the extent that we're phobic about sadness, we will feel worse. But loss, when experienced wholeheartedly, can lead to a new sense of aliveness, and to a re-engagement with the outside world. When we're in a depressed state, we often feel numb or deadened to our emotions. Depression is a chronic mental illness that needs help. Sadness, on the other hand, can be an awakening. There is a freedom in sadness that is impossible to conceive of when we're all consumed with the urge *not* to be sad. Sadness is the temporary emotion that we all feel on occasions when we've been hurt or something is wrong in our lives. It's a message. But if we don't listen, it's more likely to tip into something else. So we all need to be sad, *well*. We need to keep going, together, in order to live a meaningful life.

I take a train to meet my mother and arrive to find her waiting—a vision in the red beret she has taken to wearing at every opportunity in her retirement (for which I applaud her). We walk to her car, also red, my mother humming a tune that isn't instantly recognizable. She seems nervous, exuding a strained jauntiness.

"Train journey all right?"

"Fine, thanks. How was your vacation?"

"Wonderful, thanks! We went to the famine museum, then had a really lovely tea and cake after." "We" is my mother and her new husband. Against all odds in this crazy world, she met someone and fell in love again and decided she wanted to spend the rest of her life with him. I'm happy for her.

"Sounds lovely," I say.

She begins to inspect me.

"Have you been cutting your own hair?"

"No, why?" I smooth it down, affronted. Admittedly it hasn't seen a comb this morning . . .

She examines a strand. "Again?"

I last cut my own hair when I was six years old. The "again" feels a little rich.

"No!" I swat her away and get in the car.

We drive in silence to the cemetery we have passed many times before. It's next to the sports field, where, for one winter during my teenage years, I spent *most* weekends lusting after a boy in the year above me who played something or other on a local team (it was very cold, I learned nothing of the rules of whatever game it was he was playing and, alas, the lust remained resolutely one-sided). We park, step outside and feel overwhelmed by the multiplicity of life, near infinite. As far as the eye can see in three directions, there are only gravestones, but somewhere out there, we know, is "our" grave. The scale of sorrow feels both unique and universal and I experience a deep apprehension. But instead of pushing this feeling away, I try to ride it, now. It's painful. But I do it. And it passes.

Trees stand, waiting, but there's no one else around. And we are in no rush. So we take our time.

"When was the last time you were here?" I ask.

She estimates that it was when we buried my maternal grandmother. I have no recollection of the cemetery part of this momentous event, but then, it's been a very long time. Nearly thirty years, in fact.

"I don't know why I haven't been back," she starts but then stops. We both know why: because coming here felt too raw. We're here now, though. Together. We follow the signpost to the children's burial area. A sadder place there may not be.

"Do you remember . . ." I start, not knowing quite how to ask, "where . . .?"

My mother pinches the bridge of her nose and shakes her head.

"I think," she says slowly, "there was maybe, a bush?"

But that was decades ago; anything bush-like is now a tree. We move along rows of bright white headstones, untroubled by mildew and surrounded by fresh flowers. New losses, their graves tended with care. After an hour or so, the plots become more overgrown, and we scan the names on squares of gray granite in silence for the next half an hour. Small grass markers lie flush against the earth, becoming increasingly unkempt and strewn with brambles or flowering weeds. Rather than finding this upsetting, there's something reassuring about these losses re-wilding—overrun by nature. I read about a single mother who lost a boy called Thomas. A girl called Alice who died the same year as my sister. And then, among particularly verdant clumps of moss, I see a letter "S." The rest of the name is covered up, the earth having reclaimed it long ago. I drop to my knees and pull at the moss. This comes off easily in clumps to reveal an "o" and then a "p."

"I think I've found Sophie," I tell my mom.

The knots of grass covering the remaining letters are less yielding and I realize I'm going to have to dig them out. I'm going to have to dig out my sister's grave.

"I should've bought tools," my mother mutters distractedly. "I knew I should've brought tools."

How? How would you have known that?

We look around. There is nothing, no one. I try pulling at the earth, scraping at grass and moss, until fingernails are in danger of flicking off. My mother tries but her arthritis makes it impossible. *A stick? Could I use a stick?* There are no sticks. *My shoe?* I can't bring myself to kick in the direction of my sister's grave so I look in my wallet: a donor card, a debit card and a frequent-flier card. In a moment I already recognize as surreal, I draw out the blue British Airways Club Card and slowly, methodically, cut at moss and grass and earth to uncover the letters.

S. O. P. H. I.

"Where is the 'E'?"

"Oh, yes . . ."

"What 'yes'?"

"Her name."

I uncover the letter "A."

"It's an 'a'?"

My sister's name was "Sophia."

"Sophie for short," my mom tells me (math was never our family's strong suit). Age forty, I find out my sister's name. I also learn that her middle name was Russell. The name I chose for myself to mark my paternal emancipation, age eighteen. The name I have given all my children as a middle name and another thread that connects us all. I had no idea about any of this. I feel a solid sorrow in my chest as my mother stands very still, then presses the heels of her hands into her eyes. I reach out to her and she hauls me in, crushing my lungs in a colossal hug. A smudge of time passes before we disentangle and stand, feet planted, eyes stretched up now to the sun.

"I'm glad we came," she says.

"Me, too."

Neither of us is in any hurry to leave, so we sit. I look in my bag for tissues (we need them) and also find two oranges, a half-eaten peanut butter sandwich and a toy rabbit—the residue of family life and reminder of home. My home, now. I am flooded with gratitude for what I have and resolve to treat the future lightly, aware that every moment is a coin toss. But right now, we are okay.

Sitting on the grass by my sister's grave, my mother and I are "good" sad. We're allowing the sadness in and accepting that it will be here for a while, possibly forever. That's not to say that we won't be happy, too—we can be both. We're *meant* to be both. That's how this thing works.

I am overcome by the urge to take off my shoes and feel the grass between my toes, so I do. My mother does the same.

And then we eat oranges in the sunshine.

Acknowledgments

I T'S HARD TO KNOW where to begin the gratitude list for a book that's been fermenting forever. Everyone I've interacted with for the past four decades should probably be credited somewhere. But this thing wouldn't have made it out into the world without some serious help from some outstanding human beings including Louise Haines, Sarah Thickett, David Roth-Ey, Matt Clacher, Michelle Kane, Katy Archer, Sade Omeje and all at 4th Estate—thank you so, so much. And to Hilary Swanson, Anna Paustenbach, Aidan Mahony, Melinda Mullin, Julia Kent, Suzanne Quist, and all the team at HarperOne—you rock. Amy Reeve, thank you for proofreading the crazy British woman's work and taking out any stray *u*'s.

To Anna Power, for her wisdom and warmth. To TEDx and Adam Montandon for giving me the opportunity to do a talk in 2019 that helped crystallize the ideas that formed the basis of this book. To Dr. Mark Williamson of Action for Happiness for his ongoing support and for allowing me a platform to share ideas. Thanks to everyone who came or watched or listened and shared their own views on sadness. To Meik Wiking, fellow seeker, for his generosity. To the wonderfully empathetic Jane Elfer. To Tom Quinn from Beat, Ross Cormack from Winston's Wish, Jenny Ward and the rest of the team at the Lullaby Trust for their kindness and compassion and for offering a first response to those struggling today.

I'd like to thank Julia Samuel for her staggering intuition, care and generosity. Thanks to Dr. Dean Burnett for his patience in explaining neuroscience to a laywoman. To Dr. Kenneth Kendler, Dr. Esmée Hanna, Professor Trevor Harley, Dr. Lucy Johnstone and Dr. Hannah Murray for their insight and to Dr. Ad Vingerhoets for making crying cool. Dr. Yulia Chentsova-Dutton, Professor Nathaniel Herr, Professor Henrik Høgh-Olesen and Professor John Plunkett for sharing their expertise and knowledge and getting on board with a crazy one-woman crusade to change the way we approach our emotions. To Professor Robin Dunbar, who is unfailingly helpful, and to Dr. Nelson Freimer for the truly admirable, ambitious and astonishing work he's doing with the Depression Grand Challenge. To the wise Professor Peg O'Connor, whom I so enjoyed talking to (when all this is over can we play tennis sometime?) and Professor Marwa Azab for making sensitivity the new rock and roll. To Professor Jeanne Tsai, whose enlightening research should be on the school curriculum, and to Dr. Tal Ben-Shahar for teaching us how to live, better. And Mikael Odder Nielsen, Jonas, Evy and Anita Jensen, who were kind enough to let me in on the Aalborg culture vitamin course.

Professor Thomas Dixon is taking on the long overdue task of applying the filter of emotions to history, for which I salute him. Further salutes popped to the fabulous Professor Felice Jacka and Dr. Brendon Stubbs, for proving that lifestyle interventions make a difference (even if they don't make anyone any money).

The research for this book has been incredibly inspiring and I've been struck by the imagination, ideas and "against the grain thinking" that only the brave can bring to the table. The brilliant Professor Svend Brinkmann and the equally brilliant James Wallman both got me thinking. For days.

You know when you occasionally speak to people who appear to entirely have their shit together? Alex Soojung-Kim Pang and Joshua Becker are those people. Sagacious. Kind. With great hair (irrelevant? Yes. Both possessed of hair so lustrous it's worthy of note? Also "yes").

Pulling the camera back to get some perspective has been integral to my research and I'd like to extend my wholehearted appreciation to all who helped me with this, both while working on *The Atlas of Happiness* and here. Special thanks to Nompumelelo Mungi Ngomane, who not only spreads the message of *ubuntu* but lives by it, too. To Jade Sullivan, for doing me the honor of sharing her story. To Ben Saunders, whose honesty and hilarity helped me understand some of my own summit-seeking impulses. The funny and frank Matt Rudd challenged some of my own preconceptions and reminded me that equality is impossible until we have equality of emotions. Big thanks to the phenomenal Yomi Adegoke—I've been honored to share a stage and a publisher with you. And to the wonderful Ella Mills, who has already helped so many and is so warm and kind and smart.

I have been humbled and moved by how many people were willing to share their own experiences of profound sadness and vulnerability. A huge, heartfelt thank you to Marina Fogle, for letting us both cry in her kitchen. To Adam Kay, the busiest man in publishing, who took the time to talk and tell me his story of sadness and how it takes its toll. To the magnificent Bibi Lynch, for being unflinchingly honest and increasingly "fierce." To the staggeringly talented and caring John Crace, for sharing his story and for being insanely kind to a woman suffering from sea/morning sickness on a ferry back from the Isle of Wight Literary Festival in 2016 [*waves*]. To Jeremy Vine—generous, hilarious, high-octane—it was a thrill to be in his wake one Thursday afternoon. To Robin Ince for writing *I'm a Joke and So Are You* and for talking about his feelings around therapy so candidly. To Richard Clothier for breaking down old ideas of shame, and to Henry Hitchings for speaking to me on his birthday even though he was still in his pajamas and being nice even when I was mean about boarding schools. To Hannah and Charlie Lucas from the notOK App and to Jack Baxter and Ben May from the New Normal—you are doing amazing things.

And then there's the non-work-related support team. To the healthcare services in the UK and Denmark. And to essential workers everywhere for making the world better. To T, to my friends: thank you for keeping me afloat. To the three small people at home, for just *being*. To my mother, for giving me her blessing and encouragement to write "our story." Here it is. Here we are.

Notes

Introduction

1 Forgas, J.P., "Don't Worry, Be Sad! On the Cognitive, Motivational, and Interpersonal Benefits of Negative Mood," *Current Directions in Psychological Science*, 2013, 22(3), 225–32. doi:10.1177/0963721412474458.

2 Leary, M.R., "Emotional Responses to Interpersonal Rejection," *Dialogues in Clinical Neuroscience,* 2015, 17(4), 435–41.

3 Wegner, D.M., Schneider, D.J., et al., "Paradoxical Effects of Thought Suppression," *Journal of Personality and Social Psychology*, 1987, 53, 5–13.

4 From *Winter Notes on Summer Impressions*, Fyodor Dostoevsky's 1863 account of his travels in western Europe.

5 Wegner, Daniel M., *White Bears and Other Unwanted Thoughts: Suppression, Obsession, and the Psychology of Mental Control*, Guilford Press, 1994; and Wenzlaff, R.M. and Wegner, D.M., "Thought Suppression," *Annual Review of Psychology,* 2000, 51, 59–91. https://doi.org/10.1146/annurev.psych.51.1.59.

6 Gibbs, T., Pauselli, L., et al., "Mental Health Disparities Among Black Americans During COVID-10 Pandemic," *Psychiatric Times*, October 12, 2020, https://www.psychiatrictimes.com/view/mental-health-disparities-among-black-americans-during-covid-19-pandemic; Ibrahimi, S., Yusuf, K.K., et al., "COVID-19 Devastation of African American Families:

Impact on Mental Health and the Consequence of Systemic Racism," *International Journal of MCH and AIDS*, 2020, 9(3), 390–3. doi:10.21106/ijma.408; and Gur, R.E., White, L.K., et al., "The Disproportionate Burden of the COVID-19 Pandemic Among Pregnant Black Women," *Psychiatry Research*, 2020, 293, 113475. doi:10.1016/j.psychres.2020.113475.

7 Ibrahimi, S., Yusuf, K.K., et al., "COVID-19 Devastation of African American Families: Impact on Mental Health and the Consequence of Systemic Racism."

8 The Gallup 2019 Global Emotions Report, https://www.gallup.com/analytics/248906/gallup-global-emotions-report-2019.aspx.

9 Statistic at time of writing from WHO as of January 2020. See also https://www.who.int/news-room/fact-sheets/detail/depression and GBD 2017 Disease and Injury Incidence and Prevalence Collaborators, "Global, Regional, and National Incidence, Prevalence, and Years Lived with Disability for 354 Diseases and Injuries for 195 Countries and Territories, 1990–2017: A Systematic Analysis for the Global Burden of Disease Study 2017," *Lancet*, 2018, 392(10159), 1789–1858. https://doi.org/10.1016/S0140-6736(18)32279-7. This differs from the previously estimated figure of 350 million, though the higher figure is still regularly cited.

10 There are more, but these are the six most common. https://www.health.harvard.edu/mind-and-mood/six-common-depression-types.

11 Really: contact your doctor.

12 Blanchflower, David G., and Oswald, Andrew J., "Do Humans Suffer a Psychological Low in Midlife? Two Approaches (With and Without Controls) in Seven Data Sets," IZA Discussion Paper No. 10958. Available at SSRN: https://ssrn.com/abstract=3029829.

13 As reported in an interview with Blanchflower in: Rauch, J., "The Real Roots of Midlife Crisis," *Atlantic*, December 2014. https://www.theatlantic.com/magazine/archive/2014/12/the-real-roots-of-midlife-crisis/382235/.

14 Weiss, Alexander, King, James E., et al., "Evidence for a Midlife Crisis in Great Apes Consistent with the U-Shape in Human Well-Being," *Proceedings of the National Academy of Sciences of the United States of America*, December 2012, 109(49), 19949–52. doi:10.1073/pnas.1212592109.

15 Carstensen, Laura, Turan, Bulent, et al., "Emotional Experience Improves with Age: Evidence Based on Over 10 Years of Experience Sampling," *Psychology and Aging*, 2011, 26(1), 21–33. doi:10.1037/a0021285.

16 Blanchflower and Oswald, "Do Humans Suffer a Psychological Low in Midlife?"

Chapter 1: Don't Fight It

1 Seventy-two percent of parents who are married at the time of their child's death remain married to the same person. The remaining 28 percent includes 16 percent in which one spouse had died; only 12 percent of marriages had ended in divorce. Institute of Medicine, *When Children Die: Improving Palliative and End-of-Life Care for Children and Their Families*, National Academies Press, Washington, DC, 2003. https://doi.org/10.17226/10390.

2 "Marriage & Divorce," American Psychological Association, https://www.apa.org/topics/divorce-child-custody.

3 Ginott, Haim G., *Between Parent and Child*, Crown Publications, 2nd rev. edn, 2003, 27.

4 Michaud, Anne, "The Terrible Downside of Helicopter Parenting," Pioneer Press, January 28, 2015. https://www.twincities.com/2015/01/28/anne-michaud-anne-michaud-the-terrible-downside-of-helicopter-parenting/.

5 Zisook, Sidney, and Shear, Katherine, "Grief and Bereavement: What Psychiatrists Need to Know," *World Psychiatry*, 2009, 8(2), 67–74. doi:10.1002/j.2051-5545.2009.tb00217.x.

6 Erlangsen, A., Runeson, B., et al., "Association Between Spousal Suicide and Mental, Physical, and Social Health Outcomes: A Longitudinal and Nationwide Register-Based Study," *JAMA Psychiatry*, 2017, 74(5), 456–64. doi:10.1001/jamapsychiatry.2017.0226.

7 "Physical Symptoms of Grief," Marie Curie, March 31, 2020. https://www.mariecurie.org.uk/help/support/bereaved-family-friends/dealing-grief/physical-symptoms-grief.

8 Vitlic, A., Khanfer, R., et al., "Bereavement Reduces Neutrophil Oxidative Burst Only in Older Adults: Role of the HPA Axis and Immunosenescence," *Immunity & Ageing*, 2014, 11(13). doi:10.1186/1742-4933-11-13.

9 Samuel, Julia, *Grief Works: Stories of Life, Death and Surviving*, Penguin, 2017.

10 Chentsova-Dutton, Y.E., and Tsai, J.L., "Self-focused Attention and Emotional Reactivity: The Role of Culture," *Journal of Personality and Social Psychology*, 2010, 98(3), 507–19. https://doi.org/10.1037/a0018534; Tsai, J.L., and Chentsova-Dutton, Y., "Understanding Depression Across Cultures," in Gotlib, I.H., and Hammen, C.L. (eds.), *Handbook of Depression*, Guilford Press, 2002; Chentsova-Dutton, Y.E., Tsai, J.L., and Gotlib, I.H., "Further Evidence for the Cultural Norm Hypothesis: Positive Emotion in Depressed and Control European American and Asian American Women," *Cultural*

Diversity and Ethnic Minority Psychology, 2010, 16(2), 284–95. https://doi. org/10.1037/a0017562; Kitayama, S., Mesquita, B., and Karasawa, M., "Cultural Affordances and Emotional Experience: Socially Engaging and Disengaging Emotions in Japan and the United States," *Journal of Personality and Social Psychology,* 2006, 91(5), 890–903. https://doi.org/10.1037/0022 -3514.91.5.890; Uchida, Y., Townsend, S.S.M., et al., "Emotions as Within or Between People? Cultural Variation in Lay Theories of Emotion Expression and Inference," *Personality & Social Psychology Bulletin*, 2009, 35(11), 1427–39. https://doi.org/10.1177/0146167209347322.

11 Uchida, Townsend, et al., "Emotions as Within or Between People? Cultural Variation in Lay Theories of Emotion Expression and Inference."

12 Curhan, K.B., Sims, T., et al., "Just How Bad Negative Affect Is for Your Health Depends on Culture," *Psychological Science*, 2014, 25(12), 2277–80. https://doi.org/10.1177/0956797614543802.

13 Anwar, Y., "Feeling Bad About Feeling Bad Can Make You Feel Worse," Berkeley News, August 10, 2017. https://news.berkeley.edu/2017/08/10 /emotionalacceptance/.

14 Horwitz, Allan V., and Wakefield, Jerome C., *The Loss of Sadness: How Psychiatry Transformed Normal Sorrow into Depressive Disorder*, Oxford University Press, 2007.

15 Shorter, Edward, *How Everyone Became Depressed: The Rise and Fall of the Nervous Breakdown*, Oxford University Press, 2015.

16 American Psychiatric Association, *Diagnostic and Statistical Manual of Mental Disorders*, 5th edn., 2013. The full list: 1. Depressed mood. 2. Markedly diminished interest or pleasure in all, or almost all, activities most of the day, nearly every day. 3. Significant weight loss when not dieting, or weight gain, or decrease or increase in appetite nearly every day. 4. A slowing down of thought and a reduction of physical movement (observable by others, not merely subjective feelings of restlessness or being slowed down). 5. Fatigue or loss of energy nearly every day. 6. Feelings of worthlessness or excessive or inappropriate guilt nearly every day. 7. Diminished ability to think or concentrate, or indecisiveness, nearly every day. 8. Recurrent thoughts of death, recurrent suicidal ideation without a specific plan, or a suicide attempt or a specific plan for committing suicide. Anyone experiencing the latter should contact a doctor *ASAP. In the US, the National Suicide Prevention Lifeline is 1-800-273-8255. In Australia, the crisis support service Lifeline is 13 11 14. Other international suicide helplines can be found at www.befrienders.org.*

17 According to the NHS in the UK, "the new version of the *DSM* may have long-term healthcare, as well as cultural and political, implications." "Asperger's Not in DSM-5 Mental Health Manual," Nursing Times, December 11, 2012. https://www.nursingtimes.net/roles/learning-disability -nurses/aspergers-not-in-dsm-5-mental-health-manual-11-12-2012/.

18 The founders of modern CBT include Americans John B. Watson, Rosalie Rayner, Aaron T. Beck, Albert Ellis and David H. Barlow.

19 As popularized by biochemist William Frey in the 1980s—a theory that still persists today despite evidence to the contrary. Brody, J.E., "Biological Role of Emotional Tears Emerges Through Recent Studies," *New York Times*, August 31, 1982. https://www.nytimes.com/1982/08/31 /science/biological-role-of-emotional-tears-emerges-through-recent -studies.html.

20 Gračanin, Asmir, Bylsma, Lauren M., and Vingerhoets, Ad J.J.M., "Is Crying a Self-soothing Behavior?," *Frontiers in Psychology*, 2014, 5, 502. https://www.ncbi.nlm.nih.gov/pmc/articles/PMC4035568/.

21 Gračanin, Asmir, Hendriks, Michelle C.P., and Vingerhoets, Ad J.J.M., "Are There Any Beneficial Effects of Crying? The Case of Pain Perception and Mood," presented at a meeting of the International Society for Research on Emotion (ISRE), Amsterdam, July 2019.

22 Bayart, F., Hayashi, K.T., et al., "Influence of Maternal Proximity on Behavioral and Physiological Responses to Separation in Infant Rhesus Monkeys (*Macaca Mulatta*)," *Behavioral Neuroscience,* 1990, 104(1), 98–107. doi:10.1037/0735-7044.104.1.98.

23 Oaklander, M., "The Science of Crying," *Time*, March 16, 2016. https:// time.com/4254089/science-crying/.

24 Wong, Y.J., Steinfeldt, J.A., et al., "Men's Tears: Football Players' Evaluations of Crying Behavior," *Psychology of Men & Masculinity*, 2011, 12(4), 297–310. https://www.apa.org/pubs/journals/releases/men-12-4-297.pdf.

Chapter 2: Lower Expectations

1 Rutledge, Robb B., Skandali, Nikolina, et al., "A Computational and Neural Model of Momentary Subjective Well-being," *Proceedings of the National Academy of Sciences of the United States of America*, August 2014, 111(33), 12252–7. doi:10.1073/pnas.1407535111.

2 Russell, Helen, "A Week Off from Facebook? Participants in Danish Experiment Like This," *Guardian*, November 10, 2015. https://www

.theguardian.com/media/2015/nov/10/week-off-facebook-denmark-likes
-this-happiness-friends.

3 Iranzo-Tatay, Carmen, Gimeno-Clemente, Natalia, et al., "Genetic and Environmental Contributions to Perfectionism and its Common Factors," *Psychiatry Research*, 2015, 230(3). doi:10.1016/j.psychres.2015.11.020.

4 Rasmussen, Katie E., and Troilo, Jessica, "'It Has to Be Perfect!'" The Development of Perfectionism and the Family System," *Journal of Family Theory & Review*, June 2016, 8(2), 154–72. https://doi.org/10.1111/jftr.12140.

5 Curran, Thomas, and Hill, Andrew P., "Perfectionism Is Increasing Over Time: A Meta-analysis of Birth Cohort Differences from 1989 to 2016," *Psychological Bulletin*, 2019, 145(4), 410–29. doi:10.1037/bul0000138.

6 "There was no evidence of the moderating role of gender in the effect of types of perfectionism on performance expectations." Hassan, Hala, Abd-El-Fattah, Sabry, et al., "Perfectionism and Performance Expectations at University: Does Gender Still Matter?," *European Journal of Education and Psychology*, 2012, 5(2), 133–47. doi:10.1989/ejep.v5i2.97.

7 Ben-Shahar, Dr. Tal, *The Pursuit of Perfect: How to Stop Chasing Perfection and Start Living a Richer, Happier Life*, McGraw-Hill Education, 2009.

8 Wang, Y., and Zhang, B., "The Dual Model of Perfectionism and Depression among Chinese University Students," *South African Journal of Psychiatry*, 2017, 23. doi:10.4102/sajpsychiatry.v23i0.1025; also Flett, G.L., Hewitt, P.L., et al., "Perfectionism, Life Events, and Depressive Symptoms: A Test of a Diathesis-stress Model," *Current Psychology*, 1995, 14(2), 112–37. doi:10.1007/BF02686885. In fact, there are dozens.

9 Handley, A.K., Egan, S.J., et al., "The Relationships Between Perfectionism, Pathological Worry and Generalised Anxiety Disorder," *BMC Psychiatry*, 2014, 14(98). doi:10.1186/1471-244X-14-98.

10 Hewitt, Paul, Flett, Gordon, and Ediger, Evelyn, "Perfectionism Traits and Perfectionistic Self-presentation in Eating Disorder Attitudes, Characteristics, and Symptoms," *International Journal of Eating Disorders*, 1996, 18(4), 317–26. doi:10.1002/1098-108x(199512)18:4<317::aid-eat2260180404>3.0.co;2-2.

11 Hewitt, Flett, and Ediger, "Perfectionism Traits and Perfectionistic Self-presentation in Eating Disorder Attitudes, Characteristics, and Symptoms."

12 Hill, A.P., and Curran, T., "Multidimensional Perfectionism and Burnout: A Meta-analysis," *Personality and Social Psychology Review*, 2016, 20(3), 269–88. https://doi.org/10.1177/1088868315596286.

13 Martinelli, Mary, Chasson, Gregory S., et al., "Perfectionism Dimensions as Predictors of Symptom Dimensions of Obsessive-compulsive Disorder,"

Bulletin of the Menninger Clinic, 2014, 78(2), 140–59. doi:10.1521/bumc.2014 .78.2.140.

14 Egan, S., Hattaway, M., and Kane, R., "The Relationship Between Perfectionism and Rumination in Post Traumatic Stress Disorder," *Behavioural and Cognitive Psychotherapy*, 2014, 42(2), 211–23. doi:10.1017 /S1352465812001129.

15 Kempkea, Stefan, van Houdenhove, Boudewijn, et al., "Unraveling the Role of Perfectionism in Chronic Fatigue Syndrome: Is There a Distinction Between Adaptive and Maladaptive Perfectionism?," *Psychiatry Research*, April 30, 2011, 186(2–3), 373–7. https://doi.org/10.1016/j.psychres.2010.09.016.

16 Jansson-Fröjmark, Markus, and Linton, Steven J., "Is Perfectionism Related to Pre-existing and Future Insomnia? A Prospective Study," *British Journal of Clinical Psychology*, 2007, 46(1), 119–24. doi:10.1348/014466506X158824.

17 Dragos, D., Ionescu, O., et al., "Psychoemotional Features of a Doubtful Disorder: Functional Dyspepsia," *Journal of Medicine and Life*, September 15, 2012, 5(3), 260–76.

18 Fry, P.S., and Debats, D.L., "Perfectionism and the Five-factor Personality Traits as Predictors of Mortality in Older Adults," *Journal of Health Psychology*, May 2009, 14(4), 513–24. doi:10.1177/1359105309103571 PubMed PMID: 19383652.

19 Marcus Aurelius, *The Meditations*, Book Two, trans. George Long, Avon Books, 1993.

20 Epictetus, *The Enchiridion*, trans. Elizabeth Carter, 2018.

Chapter 3: Take Time. Be Kind.

1 Mary Wollstonecraft to Archibald Hamilton Rowan, April 1795, in Wollstonecraft, Mary, *The Collected Letters of Mary Wollstonecraft*, Columbia University Press, 2004, 287.

2 Kessler, R.C., Berglund, P., et al., "Lifetime Prevalence and Age-of-Onset Distributions of DSM-IV Disorders in the National Comorbidity Survey Replication," *Archives of General Psychiatry*, 2005, 62(6), 593–602. doi:10.1001 /archpsyc.62.6.593.

3 Sawyer, Susan M., Azzopardi, Peter S., et al., "The Age of Adolescence," *Lancet*, January 17, 2018, 2(3), 223–8. doi:10.1016/S2352-4642(18)30022-1.

4 Du Bois, W.E.B., *The Souls of Black Folk*, A.C. McClurg, 1903.

5 Eddo-Lodge, Reni, *Why I'm No Longer Talking to White People About Race*, Bloomsbury, 2017.

6 Hirsch, Afua, *Brit(ish): On Race, Identity and Belonging*, Vintage, 2018.

7 Akala, *Natives: Race and Class in the Ruins of Empire*, Two Roads, 2019.

8 Adegoke, Yomi, and Uviebinené, Elizabeth, *Slay in Your Lane: The Black Girl Bible*, 4th Estate, 2018.

9 Sullivan, Jade, "The Silence Was Deafening," Mother Pukka, June 2, 2020. https://www.motherpukka.co.uk/the-silence-was-deafening/.

10 Garcia, Sandra E., "Where Did BIPOC Come From?," *New York Times*, June 17, 2020. https://www.nytimes.com/article/what-is-bipoc.html.

11 Prinstein, Mitch, *Popular: The Power of Likability in a Status-Obsessed World*, Viking, 2017.

12 Wolke, Dieter, and Lereya, Suzet Tanya, "Long-term Effects of Bullying," *Archives of Disease in Childhood*, 2015, 100(9), 879–85. doi:10.1136 /archdischild-2014-306667.

13 Takizawa, Ryu, Maughan, Barbara, and Arseneault, Louise, "Adult Health Outcomes of Childhood Bullying Victimization: Evidence from a Five-Decade Longitudinal British Birth Cohort," *American Journal of Psychiatry*, 2014, 171(7), 777–84. doi:10.1176/appi.ajp.2014.13101401.

14 Glew, Gwen, and Fan, Ming-Yu, et al., "Bullying, Psychosocial Adjustment, and Academic Performance in Elementary School," *Archives of Pediatrics & Adolescent Medicine*, 2005, 159(11), 1026–31. doi:10.1001 /archpedi.159.11.1026.

15 *The Impact of Racism on Mental Health*, Synergi Collaborative Centre, March 2018. https://synergicollaborativecentre.co.uk/wp-content/uploads /2017/11/The-impact-of-racism-on-mental-health-briefing-paper-1.pdf.

16 Kwate, Naa Oyo, and Goodman, Melody, "Cross-Sectional and Longitudinal Effects of Racism on Mental Health Among Residents of Black Neighborhoods in New York City," *American Journal of Public Health*, 2014, 105, e1–e8. doi:10.2105/AJPH.2014.302243.

17 Bremner, J.D., "Traumatic Stress: Effects on the Brain," *Dialogues in Clinical Neuroscience*, 2006, 8(4), 445–61. doi:10.31887/DCNS.2006.8.4 /jbremner.

18 Kübler-Ross, Elisabeth, *On Death and Dying: What the Dying Have to Teach Doctors, Nurses, Clergy and Their Own Families*, Macmillan, 1969.

19 Jurecic, A., "Correspondence and Comments—Cautioning Health-Care Professionals: Bereaved Persons Are Misguided Through the Stages of Grief (*Omega: Journal of Death and Dying*, 74.4)," *Omega*, 2017, 75(1), 92–4. doi:10 .1177/0030222817701499. Bonanno, G.A., *The Other Side of Sadness: What the New Science of Bereavement Tells Us About Life After Loss*, Basic Books, 2010.

20 Kessler, David, and Kübler-Ross, Elisabeth, *On Grief and Grieving: Finding the Meaning of Grief Through the Five Stages of Loss*, Scribner, 2005.

21 As St. Augustine said: "Patience is the companion of wisdom." But then, he also said: "If you can manage it, you shouldn't touch your partner, except for the sake of having children." So don't just take his word for it (or, in fact, anything . . .).

22 Schnitker, Sarah A., "An Examination of Patience and Well-being," *Journal of Positive Psychology*, 2012, 7(4), 263–80. doi:10.1080/17439760.2012.697185.

23 Schnitker, Sarah A., and Emmons, Robert A., "Patience as a Virtue: Religious and Psychological Perspectives," *Research in the Social Scientific Study of Religion*, 2007, 18, 177–207.

24 Hershfield, H.E., Mogilner, C., and Barnea, U., "People Who Choose Time over Money Are Happier," *Social Psychological and Personality Science*, 2016, 7(7), 697–706. doi:10.1177/1948550616649239.

25 Schnitker and Emmons, "Patience as a Virtue."

26 A study at Indiana University Bloomington in 2007 asked ninety-six volunteers to contribute undisclosed sums of money into a fund that would then be shared out equally. They found that patient people were far more likely to pony up for the common good than their impatient cohorts. Curry, Oliver, Price, Michael, and Price, Jade G., "Patience is a Virtue: Cooperative People have Lower Discount Rates," *Personality and Individual Differences*, 2008, 44(3), 780–85. doi:10.1016/j.paid.2007.09.023.

27 The patient among us are also better equipped to handle the daily aggravations of traffic jams, lines, IT malfunctions (whatever applies). The psychologist Sarah A. Schnitker, from Baylor University in Texas, conducted a 2012 study where she gathered 389 undergraduates and essentially tried to wind them up. She evaluated how patient participants were to start with and how much participants valued patience as a virtue, then gauged their aggravation responses in a series of forty increasingly frustrating hypothetical situations. Schnitker then measured well-being via a self-esteem inventory, a satisfaction with life scale and a depression scale. She concluded that while patient people were able to take trying times in their stride, the easily exasperated became, well, *exasperated*.

28 Schnitker and Emmons, "Patience as a Virtue." Patient people are more likely to exercise self-control, be politically active and vote, according to a 2006 study. In fact, voting in a democratic election is the ultimate in delayed gratification: attempting to usher in policies that may take years to come to

fruition. That new roundabout/library/hospital may not get built during our lifetime, but we have to trust that politicians will act as they've promised (ha!) and have the patience to wait it out.

29 Stevens, Jeffrey R., and Hauser, Marc D., "Why Be Nice? Psychological Constraints on the Evolution of Cooperation," *Trends in Cognitive Sciences*, 2004, 8(2), 60–5. doi.10.1016/j.tics.2003.12.003.

30 "Ninety-six Percent of Americans Are So Impatient They Knowingly Consume Hot Food or Beverages That Burn Their Mouths, Finds Fifth Third Bank Survey," PR Newswire, January 27, 2015. https://www.prnewswire.com/news-releases/ninety-six-percent-of-americans-are-so-impatient-they-knowingly-consume-hot-food-or-beverages-that-burn-their-mouths-finds-fifth-third-bank-survey-300026261.html.

31 Schnitker, "An Examination of Patience and Well-being."

32 Roberts, Jennifer L., "The Power of Patience: Teaching Students the Value of Deceleration and Immersive Attention," *Harvard Magazine*, November–December 2013. https://harvardmagazine.com/2013/11/the-power-of-patience.

Chapter 4: Avoid Deprivation

1 Bardone-Cone, Anna M., Wonderlich, Stephen A., et al., "Perfectionism and Eating Disorders: Current Status and Future Directions," *Clinical Psychology Review*, 2007, 27(3), 384–405. doi:10.1016/j.cpr.2006.12.005.

2 Watt Smith, Tiffany, *The Book of Human Emotions*, Wellcome Collection, 2016.

3 If you are a "healthy-diet" enthusiast, and you answer "yes" to **any** of the following questions, you may be developing orthorexia nervosa:

(1) I spend so much of my life thinking about, choosing and preparing healthy food that it interferes with other dimensions of my life, such as love, creativity, family, friendship, work and school.

(2) When I eat any food I regard to be unhealthy, I feel anxious, guilty, impure, unclean and/or defiled; even to be near such foods disturbs me, and I feel judgmental of others who eat such foods.

(3) My personal sense of peace, happiness, joy, safety and self-esteem is excessively dependent on the purity and rightness of what I eat.

(4) Sometimes I would like to relax my self-imposed "good food" rules for a special occasion, such as a wedding or a meal with family or friends, but I find that I cannot. (Note: If you have a medical condition in which

it is unsafe for you to make ANY exception to your diet, then this item does not apply.)

(5) Over time, I have steadily eliminated more foods and expanded my list of food rules in an attempt to maintain or enhance health benefits; sometimes, I may take an existing food theory and add to it with beliefs of my own.

(6) Following my theory of healthy eating has caused me to lose more weight than most people would say is good for me, or has caused other signs of malnutrition such as hair loss, loss of menstruation or skin problems.

4 Terry, Annabel, Szabo, Attila, and Griffiths, Mark, "The Exercise Addiction Inventory: A New Brief Screening Tool," *Addiction Research and Theory*, 2004, 12(5), 489–99. doi:10.1080/16066350310001637363.

Individuals are asked to rate each statement from 1 (strongly disagree) to 5 (strongly agree). If an individual scores above 24 they are said to be at risk for exercise addiction.

Exercise is the most important thing in my life.

Conflicts have arisen between me and my family and/or my partner about the amount of exercise I do.

I use exercise as a way of changing my mood (e.g., to get a buzz, to escape, etc.).

Over time I have increased the amount of exercise I do in a day.

If I have to miss an exercise session I feel moody and irritable.

If I cut down the amount of exercise I do, and then start again, I always end up exercising as often as I did before.

Chapter 5: Avoid Excess

1 Sari, Y., "Commentary: Targeting NMDA Receptor and Serotonin Transporter for the Treatment of Comorbid Alcohol Dependence and Depression," *Alcoholism, Clinical and Experimental Research*, 2017, 41(2), 275–8. doi:10.1111/acer.13310.

2 Boden, J.M., and Fergusson, D.M., "Alcohol and Depression," *Addiction*, 2011, 106(5), 906–14. doi.10.1111/j.1360-0443.2010.03351.x.

3 Cordovil De Sousa Uva, M., Luminet, O., et al., "Distinct Effects of Protracted Withdrawal on Affect, Craving, Selective Attention and Executive Functions Among Alcohol-dependent Patients," *Alcohol and Alcoholism*, 2010, 45(3), 241–6. Craig, M., Pennacchia, A., et al., "Evaluation

of Un-medicated, Self-paced Alcohol Withdrawal," *PloS One*, 2011, 6(7), e22994. Potamianos, G., Meade, T.W., et al., "Randomised Trial of Community-based Centre Versus Conventional Hospital Management in Treatment of Alcoholism," *Lancet*, 1986, 328(8510), 797–9. Shaw, G.K., Waller, S., et al., "The Detoxification Experience of Alcoholic In-patients and Predictors of Outcome," *Alcohol and Alcoholism*, 1998, 33(3), 291–303. Driessen, M., Meier, S., et al., "The Course of Anxiety, Depression and Drinking Behaviours after Completed Detoxification in Alcoholics with and Without Comorbid Anxiety and Depressive Disorders," *Alcohol and Alcoholism*, 2001, 36(3), 249–55.

4 The AA questionnaire to determine whether or not you have an alcohol addiction is as follows:

Have you ever decided to stop drinking for a week or so, but only lasted for a couple of days?

Do you wish people would mind their own business about your drinking—stop telling you what to do?

Have you ever switched from one kind of drink to another in the hope that this would keep you from getting drunk?

Have you had to have a drink in the morning during the past year?

Do you envy people who can drink without getting into trouble?

Have you had problems connected with drinking during the past year?

Has your drinking caused trouble at home?

Do you ever try to get "extra" drinks at a party because you do not get enough?

Do you tell yourself you can stop drinking any time you want to, even though you keep getting drunk when you don't mean to?

Have you missed days off work because of drinking?

Do you have blackouts?

Have you ever felt that your life would be better if you did not drink?

5 Costa, R.M., "Dissociation (Defense Mechanism)," in Zeigler-Hill, V., and Shackelford. T. (eds.), *Encyclopedia of Personality and Individual Differences*, Springer, 2016.

6 Marich, Jamie, and O'Brien, Adam, "Demystifying Dissociation: A Clinician's Guide," Psychiatry and Behavioral Health Learning Network, December 4, 2018.

7 Bernstein, Eve M., and Putnam, Frank W., "Development, Reliability, and Validity of a Dissociation Scale," *Journal of Nervous and Mental Disease*, 1986, 174(12), 727–35. https://doi.org/10.1097/00005053-198612000-00004.

8 Halim, M.H. Abd, and Sabri, Farhana, "Relationship Between Defense Mechanisms and Coping Styles Among Relapsing Addicts," *Procedia— Social and Behavioral Sciences*, 2013, 84, 1829–37. doi:10.1016/j.sbspro.2013 .07.043. Benishek, Debra, and Wichowski, Harriet, "Dissociation in Adults with a Diagnosis of Substance Abuse," *Nursing Times*, 2003, 99(20), 34–6. And some fascinating research into internet addiction and dissociation: Biolcati, Roberta, Mancini, Giacomo, and Trombini, Elena, "Brief Report: The Influence of Dissociative Experiences and Alcohol/Drugs Dependence on Internet Addiction," *Mediterranean Journal of Clinical Psychology*, 2017, 5(7). Canan, Fatih, Ataoglu, Ahmet, et al., "The Association Between Internet Addiction and Dissociation Among Turkish College Students," *Comprehensive Psychiatry*, 2011, 53(5), 422–6. doi:10.1016/j.comppsych.2011.08.006.

9 O'Connor, Peg, *Life on the Rocks: Finding Meaning in Addiction and Recovery*, Central Recovery Press, 2016.

10 Kashdan, Todd, Barrett, Lisa, and McKnight, Patrick, "Unpacking Emotion Differentiation: Transforming Unpleasant Experience by Perceiving Distinctions in Negativity," *Current Directions in Psychological Science*, 2015, 24(1), 10–16. doi:10.1177/0963721414550708.

11 GA's "Are You Living with a Compulsive Gambler?" Quiz (and who doesn't love a quiz?). If you answer YES to at least six of these questions you may well be living with a compulsive gambler (or be one yourself).

Do you find yourself constantly bothered by debt collectors?

Is the person in question often away from home for long unexplained periods of time?

Do they ever lose time from work due to gambling?

Do you feel that they cannot be trusted with money?

Do they promise faithfully that they will stop gambling, beg and plead for another chance, yet gamble again and again?

Do they ever gamble longer than they intended to, until their last dollar is gone?

Do they immediately return to gambling to try to recover losses or to win more?

Do they ever gamble to get money to solve financial difficulties, or have unrealistic expectations that gambling will bring the family material comfort and wealth?

Do they borrow money to gamble with or to pay gambling debts?

Has their reputation ever suffered due to gambling, sometimes even to the extent of committing illegal acts to finance gambling?

Have you come to the point of hiding money needed for living expenses, fearing that you and the rest of the family may go without food and clothing if you do not?

Do you search their clothing, go through their wallet/purse when the opportunity presents itself or otherwise check on their activities?

Do you hide their money?

Have you noticed a personality change in them as their gambling progresses?

Do they consistently lie to cover up or deny their gambling activities?

Do they use guilt induction as a method of shifting responsibilities for their gambling onto you?

Do you attempt to anticipate their moods or try to control their life?

Do they ever suffer remorse or depression due to gambling, sometimes to the point of self-destruction?

Has the gambling ever brought you to the point of threatening to break up the family unit?

Do you feel that life together is a nightmare?

12 Sachs, Jeffrey D., *2019 World Happiness Report: Addiction and Unhappiness in America*, Chapter 7, Center for Sustainable Development, Columbia University (available to download here: https://worldhappiness.report/ed/2019/addiction-and-unhappiness-in-america/).

13 Wilkinson, Richard, and Pickett, Kate, *The Inner Level: How More Equal Societies Reduce Stress, Restore Sanity and Improve Everyone's Well-being,* Penguin, 2019.

14 Nook, E.C., Sasse, S.F., et al., "The Nonlinear Development of Emotion Differentiation: Granular Emotional Experience Is Low in Adolescence," *Psychological Science*, 2018, 29(8), 1346–57. doi:10.1177/0956797618773357.

Chapter 6: Get Mad

1 Loyola University Health System, "When a Broken Heart Becomes a Real Medical Condition," *ScienceDaily*, February 10, 2015. www.sciencedaily.com/releases/2015/02/150210130502.htm. Broken heart syndrome, also known by the less catchy terms: "stress-induced cardiomyopathy," "Takosubo's cardiomyopathy" or "transient apical ballooning syndrome" ("Is that a balloon in your pocket or have you just been dumped?" etc.).

2 Gospel of Luke, Chapter 10. Enjoy!

3 Wilson, Kimberley, *How to Build a Healthy Brain: Reduce Stress, Anxiety and Depression and Future-Proof Your Brain*, Yellow Kite, 2020.

4 Kazén, Miguel, Künne, Thomas, et al., "Inverse Relation Between Cortisol and Anger and Their Relation to Performance and Explicit Memory," *Biological Psychology*, 2012, 91(1), 28–35. doi:10.1016/j.biopsycho.2012.05.006.

5 Aarts, H., Ruys, K.I., et al., "The Art of Anger: Reward Context Turns Avoidance Responses to Anger-related Objects into Approach," *Psychological Science*, 2010, 21(10), 1406–10. doi:10.1177/0956797610384152.

6 Adam, Hajo, and Brett, Jeanne M., "Everything in Moderation: The Social Effects of Anger Depend on Its Perceived Intensity," *Journal of Experimental Social Psychology*, May 2018, 76, 12–18. doi:10.1016/j.jesp.2017.11.014.

7 Cogley, Zac, "A Study of Virtuous and Vicious Anger," in Timpe, Kevin, and Boyd, Craig (eds.), *In Virtues and Their Vices* Oxford University Press, New York, 2014, 199–224.

8 Farley, M., Golding, J.M., et al., "Comparing Sex Buyers with Men Who Do Not Buy Sex: New Data on Prostitution and Trafficking," *Journal of Interpersonal Violence*, 2017, 32(23), 3601–25. doi:10.1177/0886260515600874. Researchers investigated attitudes and behaviors among 101 men who bought sex and 101 age-, education- and ethnicity-matched men who did not buy sex. Men who bought sex also scored higher on measures of impersonal sex and hostile masculinity.

9 By the work of the famous sexologist Alfred Kinsey.

10 Lorde, Audre, "The Uses of Anger," *Women's Studies Quarterly*, 1981, 9(3), 7–10. https://academicworks.cuny.edu/wsq/509.

11 Adegoke, Yomi, and Uviebinene, Elizabeth, *Slay in Your Lane: The Black Girl Bible*, 4th Estate, 2018.

12 Rodrigues, Ana Paula, Sousa-Uva, M., et al., "Depression and Unemployment Incidence Rate Evolution in Portugal, 1995–2013: General Practitioner Sentinel Network Data," *Revista de saude publica*, November 17, 2017, 51(98). doi:10.11606/S1518-8787.2017051006675.

13 McGee, R.E., and Thompson, N.J., "Unemployment and Depression Among Emerging Adults in 12 States, Behavioral Risk Factor Surveillance System, 2010," *Preventing Chronic Disease*, 2015, 12(140451). http://dx.doi.org/10.5888/pcd12.140451.

14 Norström, F., Waenerlund, A., et al., "Does Unemployment Contribute to Poorer Health-related Quality of Life Among Swedish Adults?," *BMC Public Health*, 2019, 19(457). https://doi.org/10.1186/s12889-019-6825-y.

15 The dating website of choice in simpler, pre-Tinder times. Dating profiles are written and uploaded by friends of single people, instead of the single person themselves. Marginally less cringeworthy.

Chapter 7: Shake Off Shame

1 No snickering at the back: this is far less rude and exciting than it sounds. Human chorionic gonadotropin (HCG) is a hormone that supports the normal development of an egg in a woman's ovary, and stimulates the release of the egg during ovulation.

2 Thoma, Marie E., McLain, Alexander C., et al., "Prevalence of Infertility in the United States as Estimated by the Current Duration Approach and a Traditional Constructed Approach," *Fertility and Sterility*, 2013, 99(5), 1324–31.e1. https://doi.org/10.1016/j.fertnstert.2012.11.037.

3 According to the National Infertility Association, https://resolve.org /infertility-101/what-is-infertility/fast-facts/.

4 Payne, Nicky, and van den Akker, Olga, "Fertility Network UK Survey on the Impact of Fertility Problems," October 2016. https://fertilitynetworkuk .org/wp-content/uploads/2016/10/SURVEY-RESULTS-Impact-of-Fertility -Problems.pdf.

5 Levine, Hagai, Jørgensen, Niels, et al., "Temporal Trends in Sperm Count: A Systematic Review and Meta-regression Analysis," *Human Reproduction Update*, November–December 2017, 23(6), 646–59. https://doi.org/10.1093 /humupd/dmx022.

6 Hanna, Esmée, and Gough, Brendan, "The Social Construction of Male Infertility: A Qualitative Questionnaire Study of Men with a Male Factor Infertility Diagnosis," *Sociology of Health & Illness*, 2019, 42(3), 465–80. doi:10.1111/1467-9566.13038.

7 Gruenewald, Tara, Kemeny, Margaret, et al., "Acute Threat to the Social Self: Shame, Social Self-esteem, and Cortisol Activity," *Psychosomatic Medicine*, 2004, 66(6), 915–24. doi:10.1097/01.psy.0000143639.61693.ef.

8 Sznycer, Daniel, Tooby, John, et al., "Shame Closely Tracks the Threat of Devaluation by Others, Even Across Cultures," *Proceedings of the National Academy of Sciences of the United States of America*, 2016, 113(10), 2625–30. doi:10.1073/pnas.1514699113.

9 Ashby, Jeffrey, Rice, Kenneth, and Martin, James, "Perfectionism, Shame, and Depressive Symptoms," *Journal of Counseling & Development*, 2006, 84(2), 148–56. doi:10.1002/j.1556-6678.2006.tb00390.x.

10 "Miscarriage," Mayo Clinic. https://www.mayoclinic.org/diseases-conditions /pregnancy-loss-miscarriage/symptoms-causes/syc-20354298.

11 Formerly the *Primary Care Companion to the Journal of Clinical Psychiatry*, this is a web-based, peer-reviewed, abstracted publication that seeks to advance the

clinical expertise of primary care physicians and other healthcare professionals who treat patients suffering from mental and neurological illnesses. Nynas, J., Narang, P., et al., "Depression and Anxiety Following Early Pregnancy Loss: Recommendations for Primary Care Providers," *Primary Care Companion for CNS Disorders*, 2015, 17(1). doi:10.4088/PCC.14r01721.

12 Freeman, Hadley, "Women Aren't Meant to Talk About Miscarriage. But I've Never Been Able to Keep a Secret," *Guardian*, May 13, 2017. https://www.theguardian.com/lifeandstyle/2017/may/13/hadley-freeman-miscarriage-silence-around-it.

13 Kadkhodai, Christen Decker, "'There Was No Child, I Told Myself': Life and Marriage After Miscarriage," *Guardian*, July 16, 2016. https://www.theguardian.com/lifeandstyle/2016/jul/16/miscarriage-pregnancy-motherhood-loneliness.

14 Hinds, Andy, "Messages of Shame Are Organized Around Gender," *Atlantic*, April 26, 2013. https://www.theatlantic.com/sexes/archive/2013/04/messages-of-shame-are-organized-around-gender/275322.

15 Mahalik, James, Morray, Elisabeth, et al., "Development of the Conformity to Feminine Norms Inventory," *Sex Roles*, 2005, 52, 417–35. doi:10.1007/s11199-005-3709-7.

16 "WHO Warns of Surge of Domestic Violence as COVID-19 Cases Decrease in Europe," United Nations, July 5, 2020. https://unric.org/en/who-warns-of-surge-of-domestic-violence-as-covid-19-cases-decrease-in-europe.

17 "Alexithymia," *Merriam-Webster.com Dictionary*, Merriam-Webster. https://www.merriam-webster.com/medical/alexithymia. Accessed May 6, 2021.

18 For more, see: Guvensel, O., "The Relationship Among Normative Male Alexithymia, Gender Role Conflict, Men's Non-romantic Relationships with Other Men, and Psychological Well-being," Dissertation, Georgia State University, 2016. Karakis, E., and Levant, R., "Is Normative Male Alexithymia Associated with Relationship Satisfaction, Fear of Intimacy and Communication Quality Among Men in Relationships?," *Journal of Men's Studies*, 2012, 20(3), 179–86. Mattila, A., "Alexithymia in Finnish General Population," Doctoral dissertation, 2009. Miles, J., "Why Do Men Struggle to Express their Feelings?" Welldoing.org, September 13, 2017. Rodman, S., "Alexithymia: 'Does My Partner Feel Anything?'," *Huffington Post*, December 6, 2017. Schexnayder, C., "The Man Who Couldn't Feel," *Brain World*, July 3, 2019. Thompson, J., "Normative Male Alexithymia," *In Search of Fatherhood*, January 2, 2010.

19 Karakis and Levant, "Is Normative Male Alexithymia Associated with

Relationship Satisfaction, Fear of Intimacy and Communication Quality Among Men in Relationships?"

Chapter 8: Stop Apologizing for Feeling

1 Barnes, Julian, *Levels of Life*, Jonathan Cape, 2013.
2 Peskin, Harvey, "Who Has the Right to Mourn?: Relational Deference and the Ranking of Grief," *Psychoanalytic Dialogues*, 2019, 29(4), 477–92. doi:10.1080/10481885.2019.1632655.

Chapter 9: The Fallacy of Arrival

1 King's College London, "Prenatal Stress Could Affect Baby's Brain," *ScienceDaily*, October 8, 2019. www.sciencedaily.com/releases/2019/10/191008094309.htm.
2 European College of Neuropsychopharmacology, "Children of Anxious Mothers Twice as Likely to Have Hyperactivity in Adolescence," *ScienceDaily*, September 9, 2019. www.sciencedaily.com/releases/2019/09/190909095021.htm.
3 By Emma Jerrold, directed by Sam Swainsbury, who, FYI, is flipping brilliant in *Mum* on BBC. Many thanks.
4 Wolke, D., Bilgin, A., and Samara, M., "Systematic Review and Meta-analysis: Fussing and Crying Durations and Prevalence of Colic in Infants," *Journal of Pediatrics*, 2017, 185, P55–61. doi:10.1016/j.jpeds.2017.02.020.
5 University of Liverpool, "Mothers Are Made to Feel Guilty Whether they Breastfeed or Formula Feed Their Baby," *ScienceDaily*, November 16, 2016. www.sciencedaily.com/releases/2016/11/161116101900.htm.
6 According to researchers for Silentnight bed company. "Parents of Newborns Miss Out on SIX MONTHS Worth of Sleep in Their Child's First Two Years," *Daily Mail*, July 23, 2010. https://www.dailymail.co.uk/news/article-1296824/Parents-newborns-miss-SIX-MONTHS-worth-sleep-childs-years.html. So, you know, not the most rigorous of research but still pretty alarming to read when you're awake at 3 a.m. mindlessly Googling while trying to soothe an angry, non-native Danish mini Viking.
7 Richter, David, Krämer, Michael D., et al., "Long-term Effects of Pregnancy and Childbirth on Sleep Satisfaction and Duration of First-time and Experienced Mothers and Fathers," *Sleep*, April 2019, 42(4), zsz015. https://doi.org/10.1093/sleep/zsz015.
8 Gordon, A.M., and Chen, S., "The Role of Sleep in Interpersonal Conflict:

Do Sleepless Nights Mean Worse Fights?," *Social Psychological and Personality Science*, 2014, 5(2), 168–75. doi:10.1177/1948550613488952.

9 I am writing this on a Friday. I have not slept for more than five hours a night since last Saturday. T and I had an almighty row this morning when he asked me not to "overstuff" a kitchen drawer at the precise moment I was cleaning up poo (not mine). Fun fact.

10 Gaynes, B.N., Gavin, N., et al., "Perinatal Depression: Prevalence, Screening Accuracy, and Screening Outcomes," Rockville: Agency for Healthcare Research and Quality (AHRQ), Evidence Report/Technology Assessment No. 119, 2005.

11 Hagen, Edward H., "The Functions of Postpartum Depression," *Evolution and Human Behavior*, September 1999, 20(5), 325–59.

12 Perry, Bruce D., *Born to Love: Why Empathy Is Essential—and Endangered*, Harper Paperbacks, 2011.

13 Stern, Daniel N., *The Motherhood Constellation: A Unified View of Parent-Infant Psychotherapy*, Routledge, 1995.

14 Dolan, Paul, *Happy Ever After: Escaping the Myths of the Perfect Life*, Allen Lane, 2019.

15 As reported in Cain, S., "Women Are Happier Without Children or a Spouse, Says Happiness Expert," *Guardian*, May 25, 2019. https://www.theguardian.com/lifeandstyle/2019/may/25/women-happier-without-children-or-a-spouse-happiness-expert.

16 Vedantam, S., host, "You 2.0: When Did Marriage Become So Hard?," Hidden Brain, NPR, August 13, 2018. https://www.npr.org/2018/08/13/638202813/you-2-0-when-did-marriage-become-so-hard via Finkel, Eli J., *The All-or-Nothing Marriage: How the Best Marriages Work*, Dutton, 2017.

17 According to research from Bupa: "Life's Big Milestones Can Negatively Impact Your Mental Health," Bupa, May 14, 2019. https://www.bupa.com/newsroom/news/lifes-milestones-can-impact-mental-health.

18 That's the latest estimate we have of the likelihood of married people getting divorced over the course of their lifetime from the Office for National Statistics—the National Archives. https://webarchive.nationalarchives.gov.uk/20160106011951/ and "What Percentages of Marriages End in Divorce?," National Archives, February 9, 2013. http://www.ons.gov.uk/ons/rel/vsob1/divorces-in-england-and-wales/2011/sty-what-percentage-of-marriages-end-in-divorce.html.

19 According to figures from the US Centers for Disease Control and Prevention as reported in Bielski, Z., "When It Comes to Marriage, the

Third Time's Not Often the Charm," *Globe and Mail*, September 29, 2016. https://www.theglobeandmail.com/life/relationships/valentines-day/when-it-comes-to-marriage-the-third-times-not-often-the-charm/article32125001/.

20 I learned while writing *The Year of Living Danishly* (Icon, 2015) that there may be a specific "happy gene" called the 5-HTT, or the "serotonin-transporter gene," that's a major target of many drugs aimed at mood regulation. Geneticist Niels Tommerup from the Department of Cellular and Molecular Medicine at the University of Copenhagen told me that if you look at the frequency of long-form 5-HTT worldwide, the native Danish population as a whole has been shown to have higher levels of the gene. Bad luck the rest of us.

21 Parsons, George D., and Pascale, Richard T., "Crisis at the Summit," *Harvard Business Review,* March 2007.

Chapter 10: Summit Syndrome

1 Gladwell, Malcolm, *David and Goliath: Underdogs, Misfits, and the Art of Battling Giants*, Penguin, 2014.

2 Eisenstadt, J.M., "Parental Loss and Genius," *American Psychologist,* 1978, 33(3), 211–23. doi:10.1037//0003-066x.33.3.211.

3 Gottman, John M., *The Marriage Clinic: A Scientifically Based Marital Therapy*, Norton Professional Books, 1999, or read a synopsis here: https://www.gottman.com/blog/the-magic-relationship-ratio-according-science/.

4 Ponzetti, Jr., James J., "Family Beginnings: A Comparison of Spouses' Recollections of Courtship," *Family Journal*, April 1, 2005, 13(2), 132–8. https://doi.org/10.1177/1066480704271249.

5 Glass, J., Simon, R.W., and Andersson, M.A., "Parenthood and Happiness: Effects of Work-Family Reconciliation Policies in 22 OECD Countries," *AJS*, November 2016, 122(3), 886–929. doi:10.1086/688892.

6 Simon, Robin W., and Caputo, Jennifer, "The Costs and Benefits of Parenthood for Mental and Physical Health in the United States: The Importance of Parenting Stage," *Society and Mental Health*, 2018, 9(3), 296–315. doi:10.1177/2156869318786760.

7 Kahneman, Daniel, Krueger, Alan B., et al., "A Survey Method for Characterizing Daily Life Experience: The Day Reconstruction Method," *Science,* December 3, 2004, 306(5702), 1776–80. doi:10.1126/science.1103572.

8 Becker, C., Kirchmaier, I., and Trautmann, S.T., "Marriage, Parenthood and Social Network: Subjective Well-being and Mental Health in Old

Age," *PLoS ONE*, 2019, 14(7), e0218704. https://doi.org/10.1371/journal
.pone.0218704.

Chapter 11: Get Some Perspective

1 Excellent short story by Charlotte Perkins Gilman, first published in 1892, about a new mother who is encouraged to take to her bed for her nerves, despite the fact the reader can ascertain she is of perfectly sound mind. Stuck in the room with the yellow wallpaper day after day, her world shrinks so much that she eventually does lose her mind. Gilman, Charlotte Perkins, *The Yellow Wallpaper*, Simon & Brown, 2011.

2 Sutherland, D., *Raise Your Glasses,* Macdonald, 1969, 16.

3 Hewitt, Rachel, *A Revolution of Feeling: The Decade That Forged the Modern Mind*, Granta, 2017.

4 Norton, M.I., and Gino, F., "Rituals Alleviate Grieving for Loved Ones, Lovers, and Lotteries," *Journal of Experimental Psychology: General*, 2014, 143(1), 266–72. doi:10.1037/a0031772.

5 I'm buggered if I can get hold of a copy and suspect that someone long ago swiped the edition in my school's library but there's a fine Cambridge University Press book on Forster edited by the late Oxford University professor David Bradshaw for anyone wanting to read more (and why wouldn't you?). Bradshaw, D. (ed.), *The Cambridge Companion to E.M. Forster*, Cambridge University Press, 2007. doi:10.1017/CCOL0521834759.

6 From FactCheck Q&A: How Posh is Parliament? https://www.channel4 .com/news/factcheck/factcheck-qa-how-posh-is-parliament.

7 Thirty-two of the fifty-five UK prime ministers at time of writing attended boarding school with twenty attending Eton alone. https://en.wikipedia .org/wiki/List_of_prime_ministers_of_the_United_Kingdom_by _education.

8 Monbiot, G., "Boarding Schools Warp Our Political Class—I Know Because I Went to One," *Guardian*, November 7, 2019. https://www .theguardian.com/commentisfree/2019/nov/07/boarding-schools-boris -johnson-bullies.

9 Schaverien, Joy, "Boarding School Syndrome: Broken Attachments a Hidden Trauma," *British Journal of Psychotherapy*, 2011, 27(2), 138–55. doi:10.1111/j.1752-0118.2011.01229.x.

10 I will never forget the very "grand" grandmother of my first boyfriend from college asking him, straight-faced, "And how many stiffies have you had this

term?" I flushed, the color of beet. Boyfriend answered: "Six." Grandmother wrinkled her nose. I wished, very much, that the parquet floor would swallow me up. Boyfriend then explained that his grandmother was referring to heavy stock invitation cards for formal birthday celebrations. Not, to be clear, erections. We all learned something that day . . .

11 Churchill, Winston, *My Early Life: A Roving Commission*, Thornton Butterworth, 1930—or Pocket Books, 1996.

12 Toye, Richard, *Churchill's Empire: The World That Made Him and the World He Made*, Henry Holt, 2010. See also Heyden, Tom, "The 10 Greatest Controversies of Winston Churchill's Career," *BBC News Magazine*, January 26, 2015, for a neat overview.

13 Dixon, Thomas, *Weeping Britannia: Portrait of a Nation in Tears*, Oxford University Press, 2015, fig. 38.

14 But still, not everyone was on board. Dixon points out how the newspaper columnist Richard Littlejohn opposed a two-minute silence for the first anniversary of the death of Princess Diana in 1998, describing the previous year's mourning as "menacing mass hysteria" and "a revolting orgy of emotional incontinence." He's nice like that.

Chapter 12: Get More Perspective

1 Russell, Helen, *The Atlas of Happiness: The Global Secrets of How to Be Happy*, Two Roads, 2018.

2 Norton, M.I., and Gino, F., "Rituals Alleviate Grieving for Loved Ones, Lovers, and Lotteries," *Journal of Experimental Psychology: General*, 2014, 143(1), 266–72. doi:10.1037/a0031772.

3 Durkheim, Émile, *The Elementary Forms of Religious Life*, 1912.

4 Evans, Stephen, "The Employees Shut Inside Coffins," *BBC News*, Seoul, December 14, 2015. https://www.bbc.com/news/magazine-34797017.

5 Brown, A., Scales, U., et al., "Exploring the Expression of Depression and Distress in Aboriginal Men in Central Australia: A Qualitative Study," *BMC Psychiatry*, August 1, 2012, 12(97). doi:10.1186/1471-244X-12-97.

6 *The Health and Welfare of Australia's Aboriginal and Torres Strait Islander Peoples*, Australian Institute of Health and Welfare, 2015.

7 The haka composed by Dr. Ken Kennedy, Koro Tini and Jamus Webster: *Haka Koiora*—**Haka for life.** *Paiahahā, Paiahahā* (**Attention! Attention!**) *He aha rā ka tāpaea ngā mahi kikino* (**Why do we wait for something bad to happen?**) *Ki te kūkūtia tātou katoa e?* (**To eventually come together?**) *Ia ha*

ha! E oho, kia tika rā (**Wake up, be true!**) *Unuhia ngā here o te kino* (**Strip away bad things like**) *Whakatakē, whakaparahako e* (**Negativity and belittling others**) *Ko te pūtake o te whakaaro, he kaikir* (**because the underlining factor is racism**) *Takatakahia Hi* (**Stomp on it**) *Wherawherahia Hi* (**Get rid of it**) *Kia tū te tangata koia anake* (**So all that remains is your true person**) *Ko au, Ko koe, ko koe, ko au, ko tāua e* (**I am you, you are me, this is us**) *Ko te mea nui o te ao* (**The greatest thing in this world**) *He Tangata, He Tangata, He Tangata e* (**'Tis people, 'Tis people, 'Tis people**).

8 Ngomane, Nompumelelo Mungi, *Everyday Ubuntu: Living Better Together, the African Way*, Bantam Press, 2019.

9 The most serious incident of aggression in the year that Lutz spent studying the islanders in the late 1970s involved the "touching of one man's shoulder by another, a violation that resulted in the immediate payment of a severe fine." From Lutz, Catherine, "The Domain of Emotion Words on Ifaluk," *American Ethnologist*, 1982, 9(1), 113–28. *JSTOR*, www.jstor.org/stable/644315.

10 Kundera, Milan, *The Book of Laughter and Forgetting*, Faber and Faber, 2001.

11 "Participants Ease Stress Levels at Crying Events," *Japan Times*, June 22, 2013. https://www.japantimes.co.jp/news/2013/06/22/national/participants-ease -stress-levels-at-crying-events.

12 Webb, Emily, "The 'Handsome Weeping Boys' Paid to Wipe Away Your Tears," *Outlook*, BBC World Service, August 25, 2016. https://www.bbc.com /news/magazine-37178014.

13 The Russian word *toska* is best explained by Vladimir Nabokov as: "a sensation of great spiritual anguish, often without any specific cause. At less morbid levels it is a dull ache of the soul, a longing with nothing to long for, a sick pining, a vague restlessness, mental throes, yearning. In particular cases it may be the desire for somebody of something specific, nostalgia, love-sickness. At the lowest level it grades into ennui, boredom." *Dusha naraspashku* or "unbuttoned soul": https://www.goodreads.com/ quotes/309633-toska---noun-t--sk---russian-word-roughly-translated-as.

14 Barto, Agniya, "Bunny Poem": "Once a little scatter-brain Left poor Bunny in the rain. What could little Bunny do? He got wet just through and through."

15 Spoiler alert for Golyavkin, Viktor, *My Kind Father*, 1963.

Chapter 13: The Tipping Point

1 Chandola, T., Booker, C.L., et al., "Are Flexible Work Arrangements Associated with Lower Levels of Chronic Stress-Related Biomarkers? A

Study of 6025 Employees in the UK Household Longitudinal Study," *Sociology*, 2019, 53(4), 779–99. https://doi.org/10.1177/0038038519826014.

2 "Burn-out an 'Occupational Phenomenon': International Classification of Diseases," World Health Organization, May 28, 2019. https://www.who.int /mental_health/evidence/burn-out/en/.

3 From an interview with Lauren Laverne on *Desert Island Discs* on BBC Radio 4 on December 8, 2019.

4 Waters, F., Chiu, V., et al., "Severe Sleep Deprivation Causes Hallucinations and a Gradual Progression Toward Psychosis with Increasing Time Awake," *Frontiers in Psychiatry*, 2018, 9, 303. doi:10.3389/fpsyt.2018.00303.

5 Waters, Chiu, et al., "Severe Sleep Deprivation Causes Hallucinations and a Gradual Progression Toward Psychosis With Increasing Time Awake."

6 Carlat, Daniel J., "Dr. Robert Spitzer: A Personal Tribute," *Clinical Psychiatry News*, January 11, 2016. https://www.mdedge.com/psychiatry /article/105698/dr-robert-spitzer-personal-tribute. Also worth reading: Carlat, Daniel, *Unhinged: The Trouble with Psychiatry—A Doctor's Revelations About a Profession in Crisis*, Free Press, 2010, 53–4.

7 Kessler, R.C., Berglund, P., et al., "Lifetime Prevalence and Age-of-Onset Distributions of *DSM-IV* Disorders in the National Comorbidity Survey Replication," *Archives of General Psychiatry,* June 2005, 62(6), 593–602. Kessler, R.C., Chiu, W.T., et al., "Prevalence, Severity, and Comorbidity of 12-month *DSM-IV* Disorders in the National Comorbidity Survey Replication," *Archives of General Psychiatry,* June 2005, 62(6), 617–27. For a good overview read "The Prevalence and Treatment of Mental Illness Today," *Harvard Mental Health Letter*, March 2014. https://www.health .harvard.edu/mind-and-mood/the-prevalence-and-treatment-of-mental -illness-today.

8 World Health Organization, "Mental Disorders Affect One in Four People: World Health Report," September 28, 2001. https://www.who.int /whr/2001/media_centre/press_release/en/.

9 Brown, G.W., Bhrolchain, M.N., and Harris, T., "Social Class and Psychiatric Disturbance Among Women in an Urban Population," *Sociology*, 1975, 9(2), 225–54. doi:10.1177/003803857500900203.

Chapter 14: Going Pro

1 Conway, H., and Ashley, A., "Banishing Taboos to Munch Their Pills and Be Happy," *CPH Post*, November 5, 2016. http://cphpost.dk/?p=73178.

2 According to Professor Anders Petersen at Aalborg University.

3 Jeffries, Stuart, "Happiness Is Always a Delusion," *Guardian*, July 19, 2006. https://www.theguardian.com/books/2006/jul/19/booksonhealth.healthand wellbeing.

4 Cowen, Philip, "Serotonin and Depression: Pathophysiological Mechanism or Marketing Myth?," *Trends in Pharmacological Sciences*, 2008, 29(9), 433–6. doi:10.1016/j.tips.2008.05.004.

5 Eveleigh, R., Speckens, A., et al., "Patients' Attitudes to Discontinuing Not-indicated Long-term Antidepressant Use: Barriers and Facilitators," *Therapeutic Advances in Psychopharmacology*, 2019, 9. doi:10.1177 /2045125319872344.

6 Eveleigh, Speckens, et al., "Patients' Attitudes to Discontinuing Not-indicated Long-term Antidepressant Use: Barriers and Facilitators."

7 Wood, J.V., Perunovic, W.Q. Elaine, and Lee, J.W., "Positive Self-Statements: Power for Some, Peril for Others," *Psychological Science*, 2009, 20(7), 860–66. doi:10.1111/j.1467-9280.2009.02370.x.

8 Hung, Ching-I, Liu, Chia-Yih, and Yang, Ching-Hui, "Untreated Duration Predicted the Severity of Depression at the Two-year Follow-up Point," *PLoS ONE*, 2017, 12, e0185119. doi:10.1371/journal.pone.0185119.

9 Hollis, James, *Finding Meaning in the Second Half of Life: How to Finally, Really Grow Up*, Gotham Books, 2005.

10 Samuel, Julia, *Grief Works: Stories of Life, Death and Surviving*, Penguin, 2017, 165.

11 Perry, Philippa, *The Book You Wish Your Parents Had Read (and Your Children Will Be Glad That You Did)*, Penguin Life, 2019.

Chapter 15: The Buddy System

1 "Buddy system," *Merriam-Webster.com Dictionary*, Merriam-Webster. https://www.merriam-webster.com/dictionary/buddy%20system. Accessed May 12, 2021.

2 Perry, Bruce, *The Boy Who Was Raised as a Dog: And Other Stories from a Child Psychiatrist's Notebook*, Basic Books, 2007.

3 Social anxiety disorder affects approximately 15 million American adults and is the second most commonly diagnosed anxiety disorder following specific phobia. "Social Anxiety Disorder," Anxiety & Depression Association of America. https://adaa.org/understanding-anxiety/social -anxiety-disorder.

4 Graves, Robert, *Good-bye to All That: An Autobiography*, first published 1929.

5 Gilovich, T., Medvec, V.H., and Savitsky, K., "The Spotlight Effect in Social Judgment: An Egocentric Bias in Estimates of the Salience of One's Own Actions and Appearance," *Journal of Personality and Social Psychology*, 2000, 78(2), 211–22. https://doi.org/10.1037/0022-3514.78.2.211.

6 Boothby, E.J., Cooney, G., et al., "The Liking Gap in Conversations: Do People Like Us More Than We Think?," *Psychological Science*, 2018, 29(11), 1742–56. doi:10.1177/0956797618783714.

7 National Academies of Sciences, "Social Isolation and Loneliness in Older Adults: Opportunities for the Health Care System," February 27, 2020. https://www.nap.edu/catalog/25663/social-isolation-and-loneliness-in -older-adults-opportunities-for-the-.

8 Lim, Michelle, Rodebaugh, Thomas, et al., "Loneliness over Time: The Crucial Role of Social Anxiety," *Journal of Abnormal Psychology*, 2016, 125(5), 620–30. doi:10.1037/abn0000162.

9 Conklin, Annalijn I., Forouhi, Nita G., et al., "Social Relationships and Healthful Dietary Behaviour: Evidence from Over-50s in the EPIC Cohort, UK," *Social Science and Medicine*, January 2014, 100(100), 167–75. doi:10.1016/j.socscimed.2013.08.018.

10 Holt-Lunstad, J., Smith, T.B., and Layton, J.B., "Social Relationships and Mortality Risk: A Meta-analytic Review," *PLoS Med*, 2010, 7(7), e1000316. doi:10.1371/journal.pmed.1000316.

11 Anyone experiencing similar thoughts should contact a doctor. *In the US, the National Suicide Prevention Lifeline is 1-800-273-8255. Other international suicide helplines can be found at www.befrienders.org.*

12 "Peer Support: Research and Reports," Mental Health America. https://www.mhanational.org/peer-support-research-and-reports.

13 Wellman, Barry, and Wortley, Scot, "Different Strokes from Different Folks: Community Ties and Social Support," *American Journal of Sociology*, 1990, 96(3). doi:10.1086/229572.

14 Anderson, G. Oscar, and Thayer, Colette E., *Loneliness and Social Connections: A National Survey of Adults 45 and Older*, AARP Research, September 2018. https://doi.org/10.26419/res.00246.001.

15 Russell, D., "UCLA Loneliness Scale (Version 3): Reliability, Validity, and Factor Structure," *Journal of Personality Assessment*, 1996, 66(1), 20–40. Description of measure: A twenty-item scale designed to measure one's subjective feelings of loneliness as well as feelings of social isolation. Participants rate each item on a scale from 1 (Never) to 4 (Often).

16 In the UK, men are currently entitled to shared parental leave of up to thirty-seven weeks at full pay. But in reality, many men limit the time they're home to the two-week mark, largely because of cultural norms and economic pressures, since—yawn—men still outearn women in nearly every industry going (STILL! In 2021!). In fact, the UK is among the least family-friendly of the world's richest countries, according to a UNICEF assessment of policies on childcare and parental leave (*Are the World's Richest Countries Family Friendly? Policy in the OECD and EU 2019*, UNICEF report available to download as PDF via https://www.unicef-irc.org/family-friendly). Sweden, Norway, Iceland, Estonia and Portugal rank highest for family-friendly policies in OECD and EU countries. The worst? The US, where provision for parents is nonexistent and one in four women go back to work within ten days of giving birth (according to the American College of Obstetricians and Gynecologists, *ACOG Postpartum Toolkit: Returning to Work and Paid Leave*, available to download as a PDF via https://www.acog.org). At the time of writing, the US is one of just two countries without a mandatory paid leave policy. By contrast, Estonia offers women eighty-five weeks' maternity leave at full pay after having a baby.

17 Moss, R., "Samaritans Teams Up with Hairdressers and Barbers to Highlight the Life-saving Power of Listening," *Huffington Post*, July 24, 2017. https://www.huffingtonpost.co.uk/entry/samaritans-teams-up -with-hairdressers-to-highlight-the-life-saving-power-of-listening_uk _5975b8b5e4b0e79ec19a6125?guccounter=1.

18 "Suicide: One Person Dies Every 40 Seconds," World Health Organization, September 9, 2019. https://www.who.int/news-room/detail/09-09-2019 -suicide-one-person-dies-every-40-seconds.

19 US Centers for Disease Control and Prevention (CDC), *Data & Statistics Fatal Injury Report for 2018*. https://www.cdc.gov/ suicide data sheet PDF.

20 "Our Model," The Confess Project. https://www.theconfessproject.com /our-model.

21 According to the American Foundation for Suicide Prevention, "Suicide Statistics." https://afsp.org/suicide-statistics.

22 Hewings-Martin, Y., "Do All Men Die Equally," *MedicalNewsToday*, June 19, 2020. https://www.medicalnewstoday.com/articles/leading-causes -of-death-in-men.

23 Orlando Health, "Survey Finds Why Most Men Avoid Doctor Visits," ScienceDaily, June 9, 2016, www.sciencedaily.com/releases/2016/06 /160609064534.htm; Thompson, A.E., Anisimowicz, Y., et al, "The

Influence of Gender and Other Patient Characteristics on Health Care-seeking Behaviour: A QUALICOPC Study," *BMC Family Practice*, 2016, 17(38), https://doi.org/10.1186/s12875-016-0440-0; Landro, Laura, "Why Men Won't Go to the Doctor, and How to Change That," *Wall Street Journal*, April 29, 2019, https://www.wsj.com/articles/why-men-wont-go -to-the-doctor-and-how-to-change-that-11556590080; Harvard Men's Health Watch, "Mars vs. Venus: The Gender Gap in Health," Harvard Health Publishing, August 26, 2019, https://www.health.harvard.edu /newsletter_article/mars-vs-venus-the-gender-gap-in-health.

24 Cleveland Clinic Newsroom, News Releases, "Cleveland Clinic Survey: Men Will Do Almost Anything to Avoid Going to the Doctor," September 4, 2019, https://newsroom.clevelandclinic.org/2019/09/04/cleveland-clinic -survey-men-will-do-almost-anything-to-avoid-going-to-the-doctor.

25 Branson, Ken, "The Tougher Men Think They Are, the Less Likely They Are to Be Honest with Doctors," Rutgers Today, March 23, 2016, https:// www.rutgers.edu/news/tougher-men-think-they-are-less-likely-they-are -be-honest-doctors#.WRDhldLyvIV.

26 Himmelstein, Mary, and Sanchez, Diana, "Masculinity in the Doctor's Office: Masculinity, Gendered Doctor Preference and Doctor-Patient Communication," *Preventive Medicine*, 2016, 84, 34–40. doi:10.1016/j.ypmed .2015.12.008.

Chapter 16: Support Network Needed

1 Kay, Adam, *This Is Going to Hurt: Secret Diaries of a Junior Doctor*, Picador, 2017.

2 Interestingly, divorce among doctors is less common than among non-healthcare workers. Female physicians have a substantially higher prevalence of divorce than male physicians, which may be partly attributable to a differential effect of hours worked on divorce, according to a 2015 study published in Ly, D.P., Seabury, S.A., and Jena, A.B., "Divorce Among Physicians and Other Healthcare Professionals in the United States: Analysis of Census Survey Data," *British Medical Journal*, 2015, 350, h706. Apparently, it's the police you want to watch in the domestic disharmony stakes—thus the phrase: "Join the force, get a divorce!" (Hey, I'm just the messenger here . . .)

3 Sandstrom, G.M., and Dunn, E.W., "Is Efficiency Overrated?: Minimal Social Interactions Lead to Belonging and Positive Affect," *Social*

Psychological and Personality Science, 2014, 5(4), 437–42. https://doi.org/10
.1177/1948550613502990.

4 Small, Mario, "Weak Ties and the Core Discussion Network: Why People Regularly Discuss Important Matters with Unimportant Alters," *Social Networks*, 2013, 35(3), 470–83. doi:10.1016/j.socnet.2013.05.004.

Chapter 17: Take Your Culture Vitamins

1 Rauscher, F., Shaw, G., and Ky, C., "Music and Spatial Task Performance," *Nature*, 1993, 365, 611. https://doi.org/10.1038/365611a0.

2 Chang, Mei-Yueh, Chen, Chung-Hey, and Huang, Kuo-Feng, "Effects of Music Therapy on Psychological Health of Women During Pregnancy," *Journal of Clinical Nursing*, 2008, 17(19), 2580–87. doi:10.1111/j.1365-2702 .2007.02064.x.

3 CBA recipients of a B6 cardiac graft that were exposed to opera music and Mozart had significantly prolonged allograft survival (median survival times [MSTs] 26.5 and 20 days, respectively), whereas those exposed to a single sound frequency (100, 500, 1,000, 5,000, 10,000, or 20,000 Hz) or Enya did not (MSTs 7.5, 8, 9, 8, 7.5, 8.5 and 11 days, respectively).

4 Uchiyama, M., Jin, X., et al., "Auditory Stimulation of Opera Music Induced Prolongation of Murine Cardiac Allograft Survival and Maintained Generation of Regulatory CD4$^+$CD25$^+$ Cells," *Journal of Cardiothoracic Surgery*, 2012, 7, 26. https://doi.org/10.1186/1749-8090-7-26.

5 Millgram, Y., Joormann, J., et al., "Sad as a Matter of Choice?: Emotion-Regulation Goals in Depression," *Psychological Science*, 2015, 26(8), 1216–28. https://doi.org/10.1177/0956797615583295; and Yoon, S., Verona, E., et al., "Why Do Depressed People Prefer Sad Music?," *Emotion,* 2019, 20(4), 613–24. Advance online publication. https://doi.org/10.1037/emo0000573.

6 Van den Tol, A.J.M., and Edwards, J., "Exploring a Rationale for Choosing to Listen to Sad Music When Feeling Sad," *Psychology of Music*, 2013, 41(4), 440–65. doi:10.1177/0305735611430433.

7 People in the UK, Germany, Poland, the Netherlands, France, Denmark and Sweden answered an online survey by Sonos assessing how sound affected their lives.

8 Pearce, Eiluned, Launay, Jacques, and Dunbar, Robin I.M., "The Ice-breaker Effect: Singing Mediates Fast Social Bonding," *Royal Society Open Science*, October 2015, 2(10). http://doi.org/10.1098/rsos.150221. Weinstein, Daniel A., Launay, Jacques, et al., "Singing and Social Bonding: Changes in Connectivity

and Pain Threshold as a Function of Group Size," *Evolution and Human Behavior*, March 2016, 37(2), 152–8.

9 Bolwerk, A., Mack-Andrick, J., et al., "How Art Changes Your Brain: Differential Effects of Visual Art Production and Cognitive Art Evaluation on Functional Brain Connectivity," *PLoS ONE*, 2014, 9(7), e101035. https://doi.org/10.1371/journal.pone.0101035.

10 Fancourt, Daisy, and Williamon, Aaron, "Attending a Concert Reduces Glucocorticoids, Progesterone and the Cortisol/DHEA Ratio," *Public Health*, 2016, 132, 101–4. doi:10.1016/j.puhe.2015.12.005.

11 The UK's Art Fund has a new funding program in response to COVID-19; go to artfund.org; in the US, there is https://www.arts.gov/about/nea-on-covid-19/resources-for-artists-and-arts-organizations.

12 The pen name of nineteenth-century French writer Marie-Henri Beyle who, as a forerunner of Madonna and Kylie Minogue, decided that just the one name would do him and landed on Stendhal. During a visit to Italy in 1817, Stendhal visited the Basilica of Santa Croce and did a lot of staring at Volterrano's fresco of the Sibyls. "I was already in a kind of ecstasy," he writes, "by the idea of being in Florence, and the proximity of the great men whose tombs I had just seen. Absorbed in contemplating sublime beauty, I saw it close-up—I touched it, so to speak. I had reached that point of emotion where the heavenly sensations of the fine arts meet passionate feeling. As I emerged from Santa Croce, I had palpitations (what they call an attack of the nerves in Berlin); the life went out of me, and I walked with the fear of falling." Bamforth, Iain, "Stendhal's Syndrome," *British Journal of General Practice*, 2010, 60(581), 945–6. doi:10.3399/bjgp10X544780.

13 Stickley, Theodore, and Hui, Ada, "Social Prescribing Through Arts on Prescription in a UK City: Participants' Perspectives (Part 1)," *Public Health*, July 2012, 126(7), 574–9. doi:10.1016/j.puhe.2012.04.002.

14 While social prescribing tends not to be cost neutral at the start because of set-up expenses, it provides a cost-effective strategy in the medium to longer term. Rotherham CGG projects a return on investment of £3.38 for every £1 spent after five years, according to the All-Party Parliamentary Group on Arts, Health and Wellbeing, *Creative Health: The Arts for Health and Wellbeing*, 2017. Also see McDaid, D., and Park, A., *Investing in Arts on Prescription: An Economic Perspective*, London School of Economics, 2013.

15 Brandling, J., and House, W., *Investigation into the Feasibility of a Social Prescribing Service in Primary Care: A Pilot Project*, University of Bath and Bath and North East Somerset NHS Primary Care Trust, 2007.

16 "National Arts and Health Framework," https://www.arts.qld.gov.au/images /documents/artsqld/Research/National-Arts-and-Health-Framework-May -2014.pdf.

17 Cultural Commissioning Programme, "Arts on Prescription: Arts-Based Social Prescribing for Better Mental Wellbeing," 2015, https://www .artshealthresources.org.uk/docs/arts-on-prescription-arts-based-social -prescribing-for-better-mental-wellbeing/.

18 Brinkmann, Svend, *Stå fast* [in Danish], Gyldendal Business, 2014; UK edn, Polity, 2017.

Chapter 18: Read All About It

1 Berns, Gregory S., Blaine, Kristina, et al., "Short- and Long-term Effects of a Novel on Connectivity in the Brain," *Brain Connectivity*, December 2013, 3(6), 590–600. http://doi.org/10.1089/brain.2013.0166.

2 Kidd, D., and Castano, E., "Reading Literary Fiction and Theory of Mind: Three Preregistered Replications and Extensions of Kidd and Castano (2013)," *Social Psychological and Personality Science*, 2019, 10(4), 522–31. doi:10.1177/1948550618775410.

3 de Botton, Alain, *A Velocity of Being: Letters to a Young Reader*, Enchanted Lion Books, 2019.

4 Tóibín, Colm, "Ten Rules for Writing Fiction (Part Two)," *Guardian*, February 20, 2010. https://www.theguardian.com/books/2010/feb/20/10 -rules-for-writing-fiction-part-two.

5 Ballard, J.G., *Miracles of Life: Shanghai to Shepperton: An Autobiography*, 4th Estate, 2014. Ballard was also interned in a Japanese prison camp as a teenager, then went to boarding school before studying medicine. So the rest of his life was hardly plain sailing, either. But if we're all to take the "hero of our own journey" approach, it's the work + parenting aspect that I relate to.

6 Jean in Alcott, Louisa May, *Behind a Mask or, A Woman's Power*, 1866.

7 I recommend both Tomalin, Claire, *The Life and Death of Mary Wollstonecraft*, Penguin, 1992, and Todd, Janet, *Mary Wollstonecraft: A Revolutionary Life*, Bloomsbury, 2014.

8 Noah, Trevor, *Born a Crime: Stories from a South African Childhood*, Random House, 2016.

9 Janteloven or the Law of Jante is a code of conduct known in Nordic countries characterized by not being overly flashy, not thinking you're

anything special and not standing out—being overtly personally ambitious is seen as unworthy and inappropriate. The ten rules of Jante Law were first laid out by the Danish-Norwegian author Aksel Sandemose in his 1933 novel *A Fugitive Crosses His Tracks*, but the attitudes themselves date back further. The ten rules are: you're not to think you are anything special; you're not to think you are as good as we are; you're not to think you are smarter than we are; you're not to imagine yourself better than we are; you're not to think you know more than we do; you're not to think you are more important than we are; you're not to think you are good at anything; you're not to laugh at us; you're not to think anyone cares about you; you're not to think you can teach us anything.

10 Wallman, James, *Stuffocation: Living More with Less*, Penguin, 2015.

11 The seventeen stages in Joseph Campbell's hero's journey, as outlined in Campbell, Joseph, *The Hero with A Thousand Faces*, New World Library, 3rd edn, 2012, are:

> The call to adventure
> Refusal of the call
> Supernatural aid
> Crossing the threshold
> Belly of the whale
> The road of trials
> The meeting with the goddess
> Woman as temptress
> Atonement with the father
> Apotheosis
> The ultimate boon
> Refusal of the return
> The magic flight
> Rescue from without
> The crossing of the return threshold
> Master of two worlds
> Freedom to live

12 Williams, C., "The Hero's Journey: A Mudmap for Change," *Journal of Humanistic Psychology*, 2019, 59(4), 522–39. https://doi.org/10.1177/00221 67817705499.

13 Wallman, James, *Time and How to Spend It: The 7 Rules for Richer, Happier Days*, Penguin Life, 2019.

14 Bauer, Jack, McAdams, Dan, and Pals, Jennifer, "Narrative Identity and Eudaimonic Well-being," *Journal of Happiness Studies*, 2008, 9, 81–104. doi:10.1007/s10902-006-9021-6.

Chapter 19: Get Out (& Get Active)

1 Harley, Trevor, *The Psychology of Weather*, Routledge, 2018.
2 Wei, W., Lu, J.G., et al., "Regional Ambient Temperature Is Associated with Human Personality," *Nature Human Behaviour*, 2017, 1, 890–95. https://doi.org/10.1038/s41562-017-0240-0.
3 Kingma, B., and van Marken Lichtenbelt, W., "Energy Consumption in Buildings and Female Thermal Demand," *Nature Climate Change*, 2015, 5, 1054–6. https://doi.org/10.1038/nclimate2741.
4 As reported on BBC Radio 4's *Today* program, March 13, 2013.
5 Stolberg, S.G., "White House Unbuttons Formal Dress Code," *New York Times*, January 28, 2009. https://www.nytimes.com/2009/01/29/us/politics/29whitehouse.html?_r=0.
6 Cunningham, M.R., "Weather, Mood, and Helping Behavior: Quasi Experiments with the Sunshine Samaritan," *Journal of Personality and Social Psychology,* 1979, 37(11), 1947–56. https://doi.org/10.1037/0022-3514.37.11.1947.
7 Piff, P.K., Dietze, P., et al., "Awe, the Small Self, and Prosocial Behavior," *Journal of Personality and Social Psychology,* 2015, 108(6), 883–99. https://doi.org/10.1037/pspi0000018.
8 Li, Qing, "Introduction to the Japanese Society of Forest Medicine," November 29, 2008. http://forest-medicine.com/epage01.html.
9 Farrow, M.R., and Washburn, K., "A Review of Field Experiments on the Effect of Forest Bathing on Anxiety and Heart Rate Variability," *Global Advances in Health and Medicine*, 2019, 8. doi:10.1177/2164956119848654; and Morita, E., Fukuda, S., et al., "Psychological Effects of Forest Environments on Healthy Adults: Shinrin-yoku (forest-air bathing, walking) as a Possible Method of Stress Reduction," *Public Health*, 2007, 121(1), 54–63. doi:10.1016/j.puhe.2006.05.024.
10 Li, Q., Morimoto, Kanehisa, et al., "Forest Bathing Enhances Human Natural Killer Activity and Expression of Anti-Cancer Proteins," *International Journal of Immunopathology and Pharmacology*, 2007, 20(2), 3–8. doi:10.1177/03946320070200S202.
11 Li, Q., Kobayashi, Y., et al., "Effect of Phytoncide from Trees on Human

Natural Killer Cell Function," *International Journal of Immunopathology and Pharmacology*, 2009, 22(4), 951–9. doi:10.1177/039463200902200410.

12 Twohig-Bennett, Caoimhe, and Jones, Andy, "The Health Benefits of the Great Outdoors: A Systematic Review and Meta-analysis of Greenspace Exposure and Health Outcomes," *Environmental Research*, 2018, 166, 628–37. doi:10.1016/j.envres.2018.06.030.

13 Honestly, so many. Be my guest: Capaldi, C., Dopko, R.L., and Zelenski, J., "The Relationship Between Nature Connectedness and Happiness: A Meta-analysis," *Frontiers in Psychology*, 2014. doi:10.3389/fpsyg.2014 .00976. Pearson, D.G., and Craig, T., "The Great Outdoors?: Exploring the Mental Health Benefits of Natural Environments," *Frontiers in Psychology*, 2014, 5, 1178. doi:10.3389/fpsyg.2014.01178. Bratman, G.N., Hamilton, J.P., et al., "Nature Experience Reduces Rumination and Subgenual Prefrontal Cortex Activation," *Proceedings of the National Academy of Sciences of the United States of America*, 2015(112), 28, 8567–72. doi:10.1073/pnas.1510459112. Atchley, R.A., Strayer, D.L., and Atchley, P., "Creativity in the Wild: Improving Creative Reasoning through Immersion in Natural Settings," *PLoS ONE*, 2012, 7(12), e51474. doi:10.1371/journal.pone.0051474. Zhang, J.W., Piff, P.K., et al., "An Occasion for Unselfing: Beautiful Nature Leads to Prosociality," *Journal of Environmental Psychology*, 2014, 37, 61–72. doi:10.1016/j.jenvp.2013.11 .008. Mayer, F., Frantz, C., et al., "Why Is Nature Beneficial?: The Role of Connectedness to Nature," *Environment and Behavior*, 2009, 41, 607–43. doi:10.1177/0013916508319745. Joye, Y., and Bolderdijk, J.W., "An Exploratory Study into the Effects of Extraordinary Nature on Emotions, Mood, and Prosociality," *Frontiers in Psychology*, 2014, 5, 1577. doi:10.3389 /fpsyg.2014.01577. Korpela, K.M., Pasanen, T., et al., "Environmental Strategies of Affect Regulation and Their Associations with Subjective Well-Being," *Frontiers in Psychology*, 2018, 9, 562. doi:10.3389/fpsyg.2018 .00562.

14 Engemann, Kristine, Pedersen, Carsten B., et al., "Residential Green Space in Childhood Is Associated with Lower Risk of Psychiatric Disorders from Adolescence into Adulthood," *Proceedings of the National Academy of Sciences of the United States of America*, 2019, 116(11), 5188–93. 201807504. doi:10.1073/pnas.1807504116.

15 Sandseter, E.B.H., and Kennair, L.E.O., "Children's Risky Play from an Evolutionary Perspective: The Anti-phobic Effects of Thrilling Experiences," *Evolutionary Psychology*, 2011, 9(2). doi:10.1177/147470491100900212.

16 Lukianoff, Greg, and Haidt, Jonathan, *The Coddling of the American Mind: How Good Intentions and Bad Ideas Are Setting Up a Generation for Failure*, Penguin, 2018.

17 To read the full 2018 Stress in America report, view the methodology or download graphics, visit www.stressinamerica.org.

18 Mind Share Partners, "Mental Health at Work," 2019. https://www.mindsharepartners.org/mentalhealthatworkreport.

19 Press Association, "Children Spend Only Half as Much Time Playing Outside as Their Parents Did," *Guardian*, July 27, 2016, referring to a National Trust survey.

20 Wu, Gang, Feder, Adriana, et al., "Understanding Resilience," *Frontiers in Behavioral Neuroscience*, 2013, 7, 10. doi:10.3389/fnbeh.2013.00010.

21 Gaster, S., "Urban Children's Access to their Neighborhood: Changes over Three Generations," *Environment and Behavior*, 1991, 23(1), 70–85. doi:10.1177/0013916591231004.

22 Children's Society, Good Childhood Inquiry, 2007. PDF available online: Moss, Stephen, Natural Childhood Report—National Trust, 2012. https://nt.global.ssl.fastly.net/documents/read-our-natural-childhood-repo.rt.pdf.

23 Bird, W., *Natural Fit: Can Green Space and Biodiversity Increase Levels of Physical Activity?*, RSPB, 2004.

24 White, M.P., Pahl, S., et al., "The 'Blue Gym': What Can Blue Space Do for You and What Can You Do for Blue Space?," *Journal of the Marine Biological Association of the United Kingdom*, 2016, 96(1), 5–12. doi:10.1017/S0025315415002209.

25 Kelly, Catherine, "'I Need the Sea and the Sea Needs Me': Symbiotic Coastal Policy Narratives for Human Wellbeing and Sustainability in the UK," *Marine Policy*, 2018, 97, 223–31. doi:10.1016/j.marpol.2018.03.023.

26 Wheeler, Benedict W., White, Mathew, et al., "Does Living by the Coast Improve Health and Wellbeing?," *Health & Place*, September 2012, 18(5), 1198–201. https://doi.org/10.1016/j.healthplace.2012.06.015.

27 Wheeler, White et al., "Does Living by the Coast Improve Health and Wellbeing?"

28 Ratcliffe, Eleanor, *Sleep, Mood and Coastal Walking*, National Trust, 2015. https://nt.global.ssl.fastly.net/documents/sleep-mood-and-coastal-walking---a-report-by-eleanor-ratcliffe.pdf.

29 Janský, L., Pospíšilová, D., et al., "Immune System of Cold-exposed and Cold-adapted Humans," *European Journal of Applied Physiology and Occupational Physiology*, 1996, 72, 445–50. doi:10.1007/BF00242274.

30 Tulleken, Christoffer, Tipton, Mike, et al., "Open Water Swimming as a Treatment for Major Depressive Disorder," *BMJ Case Reports*, August 21, 2018. doi:10.1136/bcr-2018-225007.

31 Shevchuk, Nikolai, "Adapted Cold Shower as a Potential Treatment for Depression," *Medical Hypotheses*, 2008, 70(5), 995–1001. doi:10.1016/j.mehy .2007.04.052.

32 ACE summary here: Green, D.J., "Can Stand-up Paddleboarding Stand Up to Scrutiny?," ACE, August 2016. https://www.acefitness.org /education-and-resources/professional/prosource/august-2016/5997/ace -sponsored-research-can-stand-up-paddleboarding-stand-up-to-scrutiny/. Schram, Ben, Hing, Wayne, and Climstein, Mike, "The Long-term Effects of Stand-up Paddle Boarding: A Case Study," *International Journal of Sports and Exercise Medicine*, 2017, 3(4). doi:10.23937/2469-5718/1510065. Willmott, Ashley G.B., Sayers, Benjamin, and Brickley, Gary, "The Physiological and Perceptual Responses of Stand-up Paddle Board Exercise in a Laboratory- and Field-setting," *European Journal of Sport Science*, 2019, 20(8), 1–21. doi:10.1080/17461391.2019.1695955. Ruess, C., Kristen, K.H., et al., "Activity of Trunk and Leg Muscles During Stand Up Paddle Surfing," *Procedia Engineering*, 2013, 60, 57–61. Schram, B., Hing, W., and Clinstein, M., "The Physiological, Musculoskeletal and Psychological Effects of Stand-up Paddle Boarding," *BMC Sports Science, Medicine and Rehabilitation*, 2016, 8(32).

33 Mackay, Graham, and Neill, James, "The Effect of 'Green Exercise' on State Anxiety and the Role of Exercise Duration, Intensity, and Greenness: A Quasi-experimental Study," *Psychology of Sport and Exercise*, 2010, 11(3), 238–45. doi:10.1016/j.psychsport.2010.01.002.

34 Thompson-Coon, Jo, Boddy, Kate, et al., "Does Participating in Physical Activity in Outdoor Natural Environments Have a Greater Effect on Physical and Mental Wellbeing than Physical Activity Indoors? A Systematic Review," *Environmental Science and Technology*, 2011, 45, 1761–72. doi:10.1021/es102947t.

35 Schuch, Felipe, Vancampfort, Davy, et al., "Exercise as a Treatment for Depression: A Meta-analysis Adjusting for Publication Bias," *Journal of Psychiatric Research*, 2016, 77, 42–51. doi:10.1016/j.jpsychires.2016.02.023.

36 Schuch, Felipe, Vancampfort, Davy, et al., "Physical Activity and Incident Depression: A Meta-analysis of Prospective Cohort Studies," *American Journal of Psychiatry*, 2018, 175(7), 631–48. doi:10.1176/appi.ajp.2018 .17111194.

37 Kandola, Aaron, Lewis, Gemma, et al., "Depressive Symptoms and

Objectively Measured Physical Activity and Sedentary Behaviour
Throughout Adolescence: A Prospective Cohort Study," *Lancet Psychiatry*,
2020, 7(3), 262–71. doi:10.1016/S2215-0366(20)30034-1.

38 From @lenadunham.

39 From an 1847 letter to his twelve-year-old niece, the daughter of his sister
Petrea Severine, called Henriette, whom he addresses as Jette. I love that he
writes to her as though she is an equal, despite the fact that she is only
twelve. Respect, SK. Kierkegaard, Søren, *The Essential Kierkegaard*,
Princeton University Press, 2000.

40 Oppezzo, Marily, and Schwartz, Daniel, "Give Your Ideas Some Legs: The
Positive Effect of Walking on Creative Thinking," *Journal of Experimental
Psychology: Learning, Memory, and Cognition*, 2014, 40(4), 1142–52. doi:10
.1037/a0036577.

Chapter 20: Get Even, Mind

1 There is a great piece on this in *Harvard Business Review*: Carmichael,
Sarah Green, "The Research Is Clear: Long Hours Backfire for People and
for Companies," *Harvard Business Review*, August 19, 2015. https://hbr.org
/2015/08/the-research-is-clear-long-hours-backfire-for-people-and-for
-companies. This was echoed by Pang in his books.

2 Gladwell, Malcolm, *Outliers: The Story of Success*, Penguin, 2009.

3 Subsequent studies seeking to replicate Ericsson's have found that there's
little to separate the "good" from the "best" musicians, with each logging
an average of about 11,000 hours practice (Macnamara, Brooke N., and
Maitra, Megha, "The Role of Deliberate Practice in Expert Performance:
Revisiting Ericsson, Krampe & Tesch-Römer (1993)," *Royal Society Open
Science*, August 2019. http://doi.org/10.1098/rsos.190327). Ericsson himself
has since spoken out about the fact that the 10,000-hour mark was not
intended to represent a skills tipping point (to use another Gladwell phrase),
just the general idea that "the more you practice, the better you'll be."

4 Jenkins, Roy, *Gladstone*, Pan, 2018.

5 Nagata, Kazuaki, "Four-day Work Week Boosted Productivity by 40%,
Microsoft Japan Experiment Shows," *Japan Times*, November 5, 2019.

6 Roy, Eleanor Ainge, "Work Less, Get More: New Zealand Firm's Four-day
Week an 'Unmitigated Success,'" *Guardian*, July 19, 2018. https://www
.theguardian.com/world/2018/jul/19/work-less-get-more-new-zealand
-firms-four-day-week-an-unmitigated-success.

7 Pang, Alex Soojung-Kim, *The Distraction Addiction*, Little Brown, 2013.

8 Blankson, Amy, "4 Ways to Help Your Team Avoid Digital Distractions," *Harvard Business Review*, July 12, 2019. https://hbr.org/2019/07/4-ways-to -help-your-team-avoid-digital-distractions.

9 Ophir, Eyal, Nass, Clifford, and Wagner, Anthony D., "Cognitive Control in Media Multitaskers," *PNAS*, September 15, 2009, 106(37), 15583–7. https://doi.org/10.1073/pnas.0903620106.

10 Becker, Mark W., Alzahabi, Reem, and Hopwood, Christopher J., "Media Multitasking Is Associated with Symptoms of Depression and Social Anxiety," *Cyberpsychology, Behavior and Social Networking*, 2013, 16(2), 132–5. doi:10.1089/cyber.2012.0291.

11 Uncapher, Melina, Thieu, Monica, and Wagner, Anthony, "Media Multitasking and Memory: Differences in Working Memory and Long-term Memory," *Psychonomic Bulletin & Review*, 2015, 23, 483–490. doi:10.3758/s13423-015 -0907-3.

12 Loh, Kep-Kee, and Kanai, Ryota, "Higher Media Multi-Tasking Activity Is Associated with Smaller Gray-Matter Density in the Anterior Cingulate Cortex," *PLoS ONE*, 2014, 9, e106698. doi:10.1371/journal.pone.0106698.

13 Bush, George, Luu, Phan, and Posner, Michael I., "Cognitive and Emotional Influences in Anterior Cingulate Cortex," *Trends in Cognitive Sciences*, 2000, 4(6), 215–22.

14 Primack, Brian, Shensa, Ariel, et al., "Use of Multiple Social Media Platforms and Symptoms of Depression and Anxiety: A Nationally-representative Study Among US Young Adults," *Computers in Human Behavior*, 2017, 69, 1–9. doi:10.1016/j.chb.2016.11.013.

15 "Screen Time Stats: Here's How Much You Use Your Phone During the Workday," *RescueTime*, March 21, 2019.

16 Taylor, Nick, "Get a Deeper Understanding of Consumer Sentiment with Emotion Analysis," Brandwatch, March 18, 2019, https://www.brandwatch .com/blog/get-a-deeper-understanding-of-consumer-sentiment-with -emotion-analysis; author interview with Alex Jones, author of *The Emoji Report for Brandwatch*, July 5, 2019.

17 Perry, Philippa, *The Book You Wish Your Parents Had Read (and Your Children Will Be Glad That You Did)*, Penguin Life, 2019, 140.

18 Berger, John, *Ways of Seeing*, Penguin, 1972.

19 Farrell, Sean, "We've Hit Peak Home Furnishings, Says Ikea Boss," *Guardian*, January 18, 2016. https://www.theguardian.com/business/2016 /jan/18/weve-hit-peak-home-furnishings-says-ikea-boss-consumerism.

20 The clothing industry alone accounts for 7 percent of global greenhouse gas emissions, almost the same as the total emissions of the EU, and the climate impact of the industry is predicted to increase by 49 percent by 2030. In February 2020, Extinction Rebellion made their opposition to London Fashion Week clear, staging a protest calling for the event to be canceled and demanding that this season's "must have" is "the continuation of life on earth." Can't argue with that.

21 Lisjak, Monika, Bonezzi, Andrea, et al., "Perils of Compensatory Consumption: Within-Domain Compensation Undermines Subsequent Self-Regulation," *Journal of Consumer Research*, 2015, 41(5), 1186–203. doi:10.1086/678902.

22 Pieters, Rik, "Bidirectional Dynamics of Materialism and Loneliness: Not Just a Vicious Cycle," *Journal of Consumer Research*, 2013, 40(4), 615–31. doi:10.1086/671564.

Chapter 21: Get Even, Body

1 Jacka, Felice, *Brain Changer: The Good Mental Health Diet*, Yellow Kite, 2019.

2 Including her PhD study, nominated the most important study in psychiatry research in 2010 by Medscape Psychiatry.

3 Including the first studies in adolescents and the first study to show a link between maternal diet, as well as early life diet, and children's emotional health. Jacka, Felice, Ystrom, Eivind, et al., "Maternal and Early Postnatal Nutrition and Mental Health of Offspring by Age 5 Years: A Prospective Cohort Study," *Journal of the American Academy of Child and Adolescent Psychiatry*, 2013, 52(10), 1038–47. doi:10.1016/j.jaac.2013.07.002.

4 Jacka, Felice, O'Neil, Adrienne, et al., "A Randomised Controlled Trial of Dietary Improvement for Adults with Major Depression (the 'SMILES' Trial)," *BMC Medicine*, 2017, 15(23). doi:10.1186/s12916-017-0791-y.

5 Firth, Joseph, Marx, Wolfgang, et al., "The Effects of Dietary Improvement on Symptoms of Depression and Anxiety: A Meta-analysis of Randomized Controlled Trials," *Psychosomatic Medicine*, 2019, 81(3), 265–80. doi:10.1097/PSY.0000000000000673.

6 Firth, Marx, et al., "The Effects of Dietary Improvement on Symptoms of Depression and Anxiety."

7 Sánchez-Villegas, A., Delgado-Rodríguez, M., et al., "Association of the Mediterranean Dietary Pattern with the Incidence of Depression: The

Seguimiento Universidad de Navarra/University of Navarra Follow-up (SUN) Cohort," *Archives of General Psychiatry*, 2009, 66(10), 1090–98. https://doi.org/10.1001/archgenpsychiatry.2009.129. The Mediterranean diet has also been shown to increase the diversity of gut bacteria: De Filippis, F., Pellegrini, N., et al., "High-level Adherence to a Mediterranean Diet Beneficially Impacts the Gut Microbiota and Associated Metabolome," *Gut*, 2016, 65(11), 1812–21. doi:10.1136/gutjnl-2015-309957.

8 Wilson, Bee, "How Ultra-processed Food Took Over Your Shopping Basket," *Guardian*, February 13, 2020. https://www.theguardian.com /food/2020/feb/13/how-ultra-processed-food-took-over-your-shopping -basket-brazil-carlos-monteiro.

9 "How Much Is Too Much? The Growing Concern over Too Much Added Sugar in Our Diets," Sugar Science, University of California, San Francisco. https://sugarscience.ucsf.edu/the-growing-concern-of-overconsumption .html#.XlzcIEN7lp8.

10 United States Department of Agriculture, Economic Research Service, "USDA Sugar Supply: Tables 51-53: US Consumption of Caloric Sweeteners," 2012. https://www.ers.usda.gov/data-products/sugar-and-sweeteners-yearbook -tables.

11 In the US, nutritional labels indicate the number of servings per container rather than per 100g (as in the EU, for example). This means that although companies making processed food are now required to list the amount of trans fat in their products, if they're listing nutritional content by serving size, they can call themselves "trans fat free" as long as they contain no more than 0.5 percent trans fats in a single serving. And a "serving size" is arbitrary and can be made as tiny as the manufacturers feel like. In other words, if we eat processed foods, we're probably still eating trans fats.

12 This is lucky since Australia, where Jacka lives, trumps the UK's five-a-day fruit and veg goal with an admirable seven servings being required of all Australians. Perhaps unsurprisingly, only 5 percent of adults are currently meeting this target, according to government data. Women were more likely to eat sufficient fruit and vegetables than men (8 percent compared with 3 percent). "Diet is a quick, cheap win," says Jacka. "We showed in one study that there was a cost saving of approximately $3,000 per participant to those in the dietary group because they lost less time out of their role at work and they saw other health professionals less often. So the savings are huge, just quite apart from the human cost."

13 Avena, Nicole, Rada, Pedro, and Hoebel, Bartley, "Evidence for Sugar

Addiction: Behavioral and Neurochemical Effects of Intermittent, Excessive Sugar Intake," *Neuroscience & Biobehavioral Reviews*, 2008, 32(1), 20–39. doi:10.1016/j.neubiorev.2007.04.019.

14 Ben-Shahar, Dr. Tal, *The Pursuit of Perfect: How to Stop Chasing Perfection and Start Living a Richer, Happier Life*, McGraw-Hill Education, 2009.

Chapter 22: Do Something for Someone Else

1 The Brand Is Female, "Dear White People: Now You Know and You Can't Pretend You Don't," Medium, June 19, 2020. https://medium.com/the -brand-is-female/dear-white-people-now-you-know-and-you-cant-pretend -you-don-t-69c4005058ec.

2 Post, S., "It's Good to Be Good: 2014 Biennial Scientific Report on Health, Happiness, Longevity, and Helping Others," *International Journal of Person Centered Medicine*, 2014, 2, 1–53.

3 Studies shows doing volunteer work makes us feel better—not just in terms of correlation but in terms of causation, too. Researchers studied happiness levels in East Germany after the fall of the Berlin Wall but before the German reunification, when volunteering was still widespread. Due to the shock of the reunification, a large portion of the infrastructure of volunteering (e.g., sports clubs associated with firms) collapsed and people randomly lost their opportunities for volunteering. Based on a comparison of the change in subjective well-being of these people and of people from the control group who had no change in their volunteer status, the hypothesis is supported that volunteering is rewarding in terms of higher life satisfaction. Meier, Stephan, and Stutzer, Alois, "Is Volunteering Rewarding In Itself?," *Economica*, 2008, 75(297), 39–59. doi:10.1111/j.1468-0335.2007.00597.x.

4 Pilkington, Pamela, Windsor, Tim, and Crisp, Dimity, "Volunteering and Subjective Well-being in Midlife and Older Adults: The Role of Supportive Social Networks," *Journals of Gerontology. Series B, Psychological Sciences and Social Sciences*, 2012, 67B(2), 249–60. doi:10.1093/geronb/gbr154. This in turn can improve our self-esteem and give us a sense of belonging and a connection with our community: Brown, Kevin, Hoye, Russell, and Nicholson, Matthew, "Self-Esteem, Self-Efficacy, and Social Connectedness as Mediators of the Relationship Between Volunteering and Well-Being," *Journal of Social Service Research*, 2012, 38(4), 468–83. doi:10.1080/01488376 .2012.687706.

Doing something for someone else also puts things into perspective. We appreciate others' suffering and being aware of acts of kindness can help us have a more positive outlook on own circumstances: Otake, K., Shimai, S., et al., "Happy People Become Happier Through Kindness: A Counting Kindnesses Intervention," *Journal of Happiness Studies*, 2006, 7(3), 361–75. doi:10.1007/s10902-005-3650-z. Kerr, S.L., O'Donovan, A., and Pepping, C.A., "Can Gratitude and Kindness Interventions Enhance Well-Being in a Clinical Sample?," *Journal of Happiness Studies*, 2015, 16(1), 17–36. doi:10.1007/s10902-013-9492-1.

5 Giving time also boosts our sense of self-efficacy and so makes us more willing to commit to future giving, despite busy schedules: Mogilner, Cassie, Chance, Zoë, and Norton, Michael, "Giving Time Gives You Time," *Psychological Science*, 2012, 23(10), 1233–8. doi:10.1177/0956797612442551.

6 Dunn, Elizabeth, Aknin, Lara, and Norton, Michael, "Prosocial Spending and Happiness: Using Money to Benefit Others Pays Off," *Current Directions in Psychological Science*, 2014, 23(1), 41–7. doi:10.1177/0963721413512503.

7 Aknin, L.B., Barrington-Leigh, C.P., et al., "Prosocial Spending and Well-being: Cross-cultural Evidence for a Psychological Universal," *Journal of Personality and Social Psychology,* 2013, 104(4), 635–52. https://doi.org/10.1037/a0031578.

8 Aknin, Lara B., Dunn, Elizabeth W., and Norton, Michael I., "Happiness Runs in a Circular Motion: Evidence for a Positive Feedback Loop Between Prosocial Spending and Happiness" *Journal of Happiness Studies*, April 2012, 13(2), 347–55.

9 Andreoni, James, "Impure Altruism and Donations to Public Goods: A Theory of Warm-Glow Giving," *Economic Journal*, 1990, 100(401), 464–77. doi:10.2307/2234133.

10 According to data from the US Census Bureau and Bureau of Labor Statistics. Grimm, Robert T., Jr., and Dietz, Nathan, "Where Are America's Volunteers? A Look at America's Widespread Decline in Volunteering in Cities and States," Research Brief: Do Good Institute, University of Maryland, 2018. https://dogood.umd.edu/sites/default/files/2019-07/Where%20Are%20Americas%20Volunteers_Research%20Brief%20_Nov%202018.pdf.

11 Luke 6:38: "Give, and it will be given to you. Good measure, pressed down, shaken together, running over, will be put into your lap. For with the measure you use it will be measured back to you." Galatians 6:9: "And let us not grow weary of doing good, for in due season we will reap, if we do

not give up." Luke 6:35: "But love your enemies, and do good, and lend, expecting nothing in return, and your reward will be great, and you will be sons of the Most High, for he is kind to the ungrateful and the evil."

12 Matthew 7:12: "So whatever you wish that others would do to you, do also to them, for this is the Law and the Prophets." Galatians 6:2: "Bear one another's burdens, and so fulfill the law of Christ."

13 Armstrong, Karen, "Let's Revive the Golden Rule," TED Global, 2009. https://www.ted.com/talks/karen_armstrong_let_s_revive_the_golden _rule?language=en. She also states: "I feel an urgency about this. If we don't manage to implement the Golden Rule globally, so that we treat all peoples, wherever and whoever they may be, as though they were as important as ourselves, I doubt that we'll have a viable world to hand on to the next generation." And that was more than a decade ago. So, you know, let's get going . . . A good start is signing and sharing the Charter for Compassion, crafted by thinkers from the three Abrahamic traditions of Judaism, Christianity and Islam, and based on the fundamental principle of the Golden Rule. www.CharterForCompassion.org.

14 Read more on this: Healy, Patrick, "The Fundamental Attribution Error: What It Is and How to Avoid It," Harvard Business School Online, June 8, 2017. https://online.hbs.edu/blog/post/the-fundamental-attribution-error.

15 Lucker, Gerald, Beane, William, and Helmreich, Robert, "The Strength of the Halo Effect in Physical Attractiveness Research," *Journal of Psychology*, 1981, 107(1), 69–75. doi:10.1080/00223980.1981.9915206.

16 For information on dates and locations, visit TheNewNormalCharity.com.

17 Becker, Joshua, *The More of Less: Finding the Life You Want Under Everything You Own*, WaterBrook, 2016.

18 Fowler, James, and Christakis, Nicholas, "Cooperative Behavior Cascades in Human Social Networks," *Proceedings of the National Academy of Sciences of the United States of America*, 2010, 107(12), 5334–8. doi:10.1073/pnas.0913149107.

19 And when we witness other people's altruistic behavior, we experience a sense of elevation, which, in turn, is more likely to encourage us to do something for someone else. Schnall, Simone, Roper, Jean, and Fessler, Daniel, "Elevation Leads to Altruistic Behavior," *Psychological Science*, 2010, 21(3), 315–20. doi:10.1177/0956797609359882.

20 Weng, Helen, Fox, Andrew, et al., "Compassion Training Alters Altruism and Neural Responses to Suffering," *Psychological Science*, 2013, 24(7), 1171–80. doi:10.1177/0956797612469537.

21 Find the pattern at https://www.spruttegruppen.dk/recipes/.